Ewa Duda

Peer Effects in Green Transformation: Leveraging Social Learning

With 18 figures

V&R unipress

National Centre for Research
and Development

Bibliographic information published by the Deutsche Nationalbibliothek
The Deutsche Nationalbibliothek lists this publication in the Deutsche Nationalbibliografie;
detailed bibliographic data are available online: https://dnb.de.

This publication was funded by the EEA / Norway Grants 2014–2021 and the state budget of
Poland through the National Centre for Research and Development under grant agreement
no. NOR/IdeaLab/GREENHEAT/0006/2020.

Reviewer: Professor, PhD hab. Aleksandra Tłuściak-Deliowska and
Professor, PhD hab. Danuta Uryga

© 2025 by Brill | V&R unipress, Robert-Bosch-Breite 10, 37079 Göttingen, Germany,
an imprint of the Brill-Group
(Koninklijke Brill BV, Leiden, The Netherlands; Brill USA Inc., Boston MA, USA; Brill Asia Pte Ltd,
Singapore; Brill Deutschland GmbH, Paderborn, Germany; Brill Österreich GmbH, Vienna, Austria)
Koninklijke Brill BV incorporates the imprints Brill, Brill Nijhoff, Brill Schöningh, Brill Fink,
Brill mentis, Brill Wageningen Academic, Vandenhoeck & Ruprecht, Böhlau and V&R unipress.
Unless otherwise stated, this publication is licensed under the Creative Commons
License Attribution-Non Commercial-No Derivatives 4.0 (see https://creativecommons.org/licenses/
by-nc-nd/4.0/) and can be accessed under DOI 10.14220/9783737017718. Any use in cases other
than those permitted by this license requires the prior written permission from the publisher.

Cover image: © Ewa Duda
Proofreading: Danuta Zgliczyńska
Printed and bound by CPI books GmbH, Birkstraße 10, 25917 Leck, Germany
Printed in the EU.

Vandenhoeck & Ruprecht Verlage | www.vandenhoeck-ruprecht-verlage.com

ISBN 978-3-8471-1771-1

Contents

Introduction 7

Chapter 1. Social Learning Theory 11
    Albert Bandura's Social Learning Theory 12
    Social Learning Theory Concepts 15

Chapter 2. Peer Effects 21
    Endogenous and Exogenous Peer Effects 22
    Active and Passive Peer Effects 26
    Positive and Negative / Desirable and Undesirable Peer Effects 28
    Contemporaneous and Lagged Peer Effects 29
    Group and Individual Peer Effects 30
    Unidirectional and Bidirectional Peer Effects 31
    Symmetric and Asymmetric Peer Effects 32

Chapter 3. Mechanisms of Peer Effects Formation: Communication Channels 33
    Diffusion of Innovation Theory 33
    Observational Learning 38
    Word-of-Mouth Effect 40

Chapter 4. Mechanisms of Peer Effects Formation: Members of Social System 47
    Opinion Leaders 47
    Peer Pressure Effects 49

Chapter 5. Application of Peer Effects: Educational Interventions 55
    Reinforcing the Word-of-Mouth Effect 55
    Word-of-Mouth to Increase the Effectiveness of Educational Interventions 57

Promotional Activities 59

Chapter 6. Modelling Peer Effects 65

Conclusions 71

References 73

Introduction

The escalating global energy crisis triggered by Russia's invasion in Ukraine (International Energy Agency, 2023) started an important debate regarding the need to accelerate the elimination of fossil fuels, which are a significant share of the heating sector. The need for energy transformation concerns large and small companies as well as residential consumers. In the present monograph I focus on the latter group of consumers, as they consume a third of total energy (Reuter et al., 2021). It contributes significantly to the impact on air quality, generating environmental, economic and social consequences.

The decision made by household residents to change a fossil fuel-based heating system for a renewable energy-based system is determined by several factors, primarily those of economic and technical nature. Barriers to accessing renewable energy sources include economic barriers such as the high cost of installation, the uncertainty of recovering the costs over the lifetime of the installation, the lack of adequate subsidy programmes, as well as technical limitations that include the construction constraints of existing buildings, technological sophistication of the installations requiring support of specialists, or concerns about the operation of the installation, its failure rate. However, educational factors should be taken into account as an integral part of the decision-making process, as the decision to invest in a renewable energy installation also depends on the potential users' level of knowledge about the energy technology, its advantages and consequences of its use (Hampton & Eckermann, 2013).

Research indicates that a significant percentage of renewable energy installations are used by consumers with high level of education, particularly those who by their professions are linked with the energy sector (Shakeel & Rajala, 2020). Awareness of the benefits of the implementation of renewable energy solutions can be a catalyst for positive change in this area. Therefore, the policy should cover educational aspects – focusing on efficiency, effectiveness and forms of learning, including social learning, as it is an important part of promoting local decarbonisation. More attention should therefore be paid to these issues.

The necessity of examining the influence of social learning on the adoption of renewable energy systems in greater depth has already been recognized, highlighting the importance of social interaction and educational behavioural interventions in this process (Gillingham & Bollinger, 2021). However, many studies are fragmented, focusing on individual components of social learning. Viewing social learning as an interdisciplinary, dynamic, multi-level and complex process is imperative for achieving environmental sustainability. This perspective not only integrates the latest research on the subject but also generates novel approaches, particularly from the domain of modelling social phenomena, including the under-recognised educational dimension. Furthermore, a significant proportion of the research is tailored to the American market, considering the unique cultural, mental, and institutional contexts that apply to it. By presenting to readers, I aim to provide a more comprehensive understanding of social learning in the green transition process. This includes not only the educational and social aspects but also on the dimension related to the mathematical modelling of this phenomenon.

The importance of this scientific initiative stems from two considerations. Firstly, the monograph can serve as a bridge between the disparate fields of engineering sciences, natural sciences and social sciences, which are inextricably linked in the context of energy transition. Over the years, I have worked at the intersection of these fields -as a mathematician by training, as an environmental engineer and as an educator. The experience I have gained in the environments associated with these disciplines shows that representatives from each field read and interpret given situations differently and have varying expectations of scientific publications. I have repeatedly encountered situations where representatives from engineering and natural sciences report difficulties in understanding scientific articles published in the social sciences. Conversely, researchers in the social sciences often struggle to comprehend publications from the sciences. These difficulties arise from the distinct conventions for writing scientific articles that vary across disciplines due to their unique characteristics. Despite these differences, it is essential for these diverse scientific communities to collaborate on initiatives for a green transformation. This monograph, aims to address the needs of various scientific groups interested in the energy transition by presenting the process of social learning within this context.

I ask for the reader's forbearance here, as the effort to address the needs of specific groups of recipients may result in some simplification of the issues discussed. My goal is to present these topics in a manner that is accessible to individuals who work in different disciplines on a daily basis.

Secondly, the energy market has recently experienced a rapid geopolitical transformation, altering the public's perception of energy security issues. Consequently, this monograph aims to facilitate an understanding of the complex and nuanced social phenomena, particularly social learning, that should be considered when implementing local energy transition policies aimed at decarbonisation.

The energy transition presented here is driven by the proliferation of modern, increasingly advanced technologies. Therefore, this study is based on diffusion of innovation theory (DOI, Rogers, 1983). This theoretical framework integrates the intersecting disciplines of education, sociology, environment, economics and energy forming the basis for further mathematical modelling using a system dynamics approach. The increasing adoption of solar-based solutions can be explained by the principles of the diffusion of innovation theory. This theory suggests that individuals are more likely to adopt innovative solutions when their friends and family also use them, generating interest and acceptance of the solution (Li et al., 2023). Consequently, the learning process spreads alongside the socio-ecological transformation.

By applying the theory of diffusion of innovations as its research framework, this study aims to determine how social learning can facilitate adoption of renewable energy solutions by individual households residents (Research Goal 1). At the same time, this study also aims to show how social learning can restrict the use of renewable energy solutions by individual households residents (Research Goal 2). A systems dynamics modelling approach (Sterman, 2001) used in the study enables a deeper understanding of the interactions between different types of social learning accompanying introduction of renewable energy solutions in individual households.

This monograph introduces readers to the concept of applying social learning theory in the field of energy policy. It may interest not only for those seeking integrated knowledge on the subject intending to introduce new or update existing social policies, but also researchers exploring the issue from educational, sociological, energy or mathematical modelling perspectives.

The first chapter of the book focuses on presenting the theoretical framework used in the research. Chapter one introduces the basics of Albert Bandura's social learning theory, outlining its definitional aspects and conceptual assumptions. The theory was originally developed to explain children's knowledge acquisition processes. However, it has also been widely applied in adult education. The second chapter of the monograph introduces a typology of peer effects drawn from the categorisation proposed by Weihua An (2011). The subsequent chapter explains the mechanisms underlying the formation of peer effects. Examples provided in both chapters illustrate the process of adopting solutions to bolster the green energy transition. The final chapter introduces modelling and analytical approaches to peer effects in the area of energy transition, with a primary focus on European examples given the research context of this monograph. This is because it is part of a research project aimed at gradually phasing out of individual fossil fuel boilers in Poland.

This monograph aims to emphasise the importance of social learning in promoting green change. It is widely recognized that even the most innovative solutions cannot succeed unless accepted by the public. Therefore, considering social factors, especially multi-level learning, is essential in planning for genuine social change, ensuring initiatives do not result in superficial transformations.

Chapter 1.
Social Learning Theory

The chapter presented here starts with the topic of adult learning, as the decision to adopt a specific household energy system is typically made by adults. Lifelong learning is a resilient, dynamic and dominant theme across the field of pedagogy particularly within the European perspective and, also in Poland. Within adult learning, there are continual changes in both theoretical positions and practices. What sets it apart is the high expectations that many organisations, entities, and individuals have of it, promoting lifelong learning at various levels. The corporate sector is actively promoting the lifelong learning development, resulting in its rapid expansion. Given the breadth of this topic, the introduction to lifelong learning presented in this chapter does not cover it comprehensively but focuses on what I consider to be the most crucial aspects.

Lifelong learning, much like other concepts in the field of education, is defined in various ways. It can be narrowly defined as "development after formal education: the continuing development of knowledge and skills that people experience after formal education and throughout their lives" (London, 2011, p. 4). It can also be considered much more broadly, as "all purposeful learning activity undertaken on an ongoing basis with the aim of improving knowledge, skills and competence" (Commission of the European Communities, 2000, p. 3). Lifelong learning is increasingly viewed as a commercial product, attracting interest from the business sector where adults are seen as potential customers. As a result the educational offer is tailored to various categories of adults, including working professionals, seniors, and those engaged in civic, professional, and cultural education, promoting a pro-social agenda. Notably, the field is vibrant with the development of significant theories and viewpoints (Belzer & Dashew, 2023).

Lifelong adults education serves as a realm for constructing theories and perspectives on education and learning. The emergence of significant theoretical positions is attributed to the adult education's departure from traditional, school-based child and adolescent education while still drawing from its foundational heritage. It formulates theories specifically tailored to adults and establishes its own unique theoretical frameworks (Clark, 2021). Andragogues aim to develop these theories to encompass the comprehensive growth of adult individuals in life education, civic education, cultural education and the enhancement of quality of life and well-being. As such, these evolving theories of learning are inherently broad, addressing a wide range of issues and applicable across various fields (Jarvis, 2004; Murtonen & Lehtinen, 2020).

The current generation faces challenges from changing legislation, economic shifts, and evolving lifestyles driven by changes in the education market, the labour market, technological advancements, and increasing environmental concerns. Therefore, new theories of adult learning are emerging that incorporate these elements, often emphasising the component of change. Various perspectives can be discussed to understand adult learning. Learning can be understood in several ways. Firstly, as an effect it refers to the end result of learning, essentially the knowledge acquired. Secondly, as a process it involves individual mental operations and specific activities that lead to planned or unplanned outcomes. Lastly, learning can be understood as a change that occurs within a person based on their own actions or interactions with the social environment, including where they live and work (Fenwick & Tennant, 2020).

Numerous theories have been developed to explain human behaviour, many of which aim to represent the process of learning. Researchers (Montrey & Shultz, 2020) emphasise the evolution of social learning from perceiving and imitating others to focusing on the outcomes of knowledge or specific skills. This evolution also includes the social experience of the process over time, its modelling and transmission, and the construction of social reality. The conceptualisation of the multiple social contexts of learning underpins many andragogical theoretical approaches. However, as Ewa Kurantowicz emphasises "whichever social learning option we choose, its precondition is always the direct or indirect presence of others" (Kurantowicz, 2012, pp. 13–14).

Albert Bandura's Social Learning Theory

The origins of this approach lie in classical social learning theory introduced by Canadian-American psychologist Albert Bandura (1977), which presents society, the relationships within it as crucial learning environments. Bandura's research was embedded in social-cognitive psychology. According to social learning theory, the acquisition of knowledge and new skills, whether intentional or incidental, can be based on observing others and then imitating or adopting their behaviour. The process of modelling learning occurs primarily through the informational function. When observing actions taking place, spectators mainly acquire symbolic representations of these actions, which they then use as guides.

In Bandura's conceptualisation, presented in Figure 1, four processes govern observational learning (Bandura, 1977).

The initial factor guiding social learning pertains to attention, which facilitates the precise comprehension of noteworthy aspects of the behaviour being modelled. Therefore, the first step in the emerging social learning process is to identify occurring motif. Among the significant determinants of attention, A. Bandura distinguishes modelling stimuli, or associational patterns. The faster assimilation of a given behavioural model is possible if it is regularly linked with a particular trigger. Another factor that enhances the ability to attract attention is the prevalence or unique character of the assimilated behaviour, setting it apart from others. Emotions are another factor that supports attention. We focus attention more easily on behaviours that have a higher emotional charge. Additionally, the characteristics of the behaviours being observed such as relevance and complexity, also influence the rate and extent of observational learning. The observer's various individual characteristics significantly influence their level of attention. Factors such as sensory capacities, level of arousal during observation, perceptual set, and past reinforcement experiences are among such factors (Bandura, 2021).

The next stage of social learning concerns the memorization process, where an individual retains what they have observed. This mechanism for storing information comprises three distinct phases, each with varying durations and encoding methods. The first phase involves the sensory register. The sensory register functions as a filter by selecting and retaining only important information that arrives within a short timeframe for further processing. Each sense (hearing, sight, touch, smell, taste) has a separate sensory register, recording and storing different stimuli. The next phase involves the working memory, which stores and manipulates information, overseeing and coordinating the memorization process. After the sensory register receives certain stimuli, the working memory processes the information by integrating previously acquired knowledge with continuously incoming sensory data. Working memory serves as the primary component of the human mental structure, acting as the mediator and connector between the other two stages of memory, namely sensory and long-term memory. The system comprises the following subsystems: the central executive, the phonological loop, the visuospatial sketchpad, and the episodic buffer (Baddeley et al., 2017).

The third stage of the social learning process occurs when an individual replicates an action they have observed and remembered previously. It is uncommon for an individual to perfectly recall something they have previously observed and remembered on their first attempt. Numerous hindrances to this process encompass both individual predispositions and personal traits, and physical capabilities. Secondly, the individual's current skill set plays a decisive role, based on the accumulated component responses. A broad approach increases one's probability of properly acquiring a new skill or retaining new knowledge. Conversely, if the requirements for acquiring particularly intricate abilities or knowledge are insufficient, it becomes necessary to enhance them through additional initial repetition, preferably with the support of informative feedback (Bandura, 2021).

The fourth stage of the social learning process is driven by motivational forces, where the consequences of acquired behaviours determine whether they will be repeated or discontinued. In the end, individuals do not incorporate the behaviours or knowledge they have acquired into their daily routine automatically. They are inclined to employ what they have learned if it yields benefits or satisfaction, however, they will refrain from using gained knowledge and skills that are detrimental or uncomfortable to them. People prefer behaviours with expected positive outcomes are preferred over consequences they wish to avoid. Due to the numerous factors influencing learning through observation or interaction with others, not all models are imitated, even if they appear highly effective (Bandura, 2021).

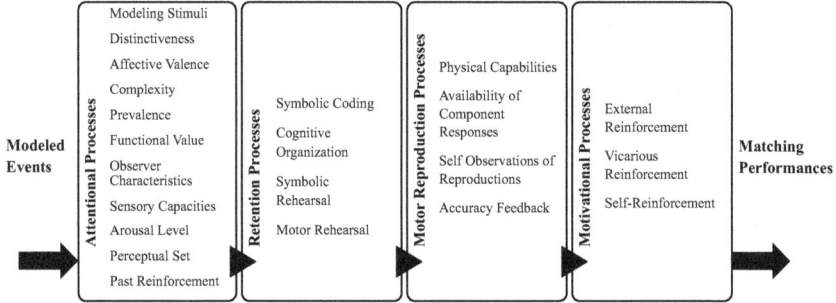

Figure 1. Component processes in the social learning. Source: Bandura, A. (1977). Social learning theory. Prentice Hall. p. 23

Following Bandura's social learning theory, it is hypothesised that how people mentally construct their experiences is crucial. The effectiveness of this learning model is enhanced by associated factors such as incentive motivation, capacity for forethought, outcome expectations (Bandura, 1999). People observe how others behave in a given situation, but they also assess their own competence to behave similarly in a similar setting and apply the adopted model to analogous circumstances. Thus, analysing observed conditions and relating them to one's own experience increases the probability of imitation and expands the range of situations in which a particular reaction is applicable. Observing others' efforts produce a desired outcome can increase the expectation that similar benefits will be achieved by putting in comparable effort. The effectiveness of incentives depends on assessments of having comparable competencies; thus utilising a corresponding behavioural model will be rewarded with similar benefits. Similarly, observing actions that result in punishment may prompt one to avoid emulating them in future (Bandura, 1999).

Another human characteristic relevant to social learning is the capacity for forethought the foreseeable future events, likely consequences of actions taken. This ability permits the formulation of strategies that can achieve intended outcomes while circumventing adverse effects. This also involves implementing and utilising actions that can lead to anticipated, specific outcomes. Individuals not only from observing the successes and failures of their peers but also from their personal experiences. People tend to adopt behaviours perceived as successful while avoiding those that have led to failure. Nonetheless, modelled behaviour is not a product of expected outcomes, but of perceived similarities. Individuals anticipate achieving similar, expected outcomes because they possess comparable predispositions towards it (Bandura, 1999).

Social Learning Theory Concepts

The classical concept of social learning theory, which relies on observing and imitating others, has evolved over time, leading to numerous developments. As a result, its parameters have broadened to encompass various aspects of human interaction and interpersonal relations, not just those focusing on cognitive processes. An analysis of selected existing models reveals various concepts of social learning, which are briefly presented in Table 1.

Table 1. Conceptual models of social learning. Author's own elaboration

Resource	Concept	Main premise
(Bandura, 1977)	Observational learning	– The significant functions of vicarious, symbolic, and self-regulatory processes in psychological functioning; – Recognition that human thought, emotions, and behaviour may be significantly shaped by observation as well as by direct experience, promoting the establishment of observational paradigms for investigating the impact of socially mediated experience; – Central role assigned to self-regulation processes.

Table 1 *(Continued)*

Resource	Concept	Main premise
(Wenger, 2000)	Communities of Practice	- Social learning in terms of personal experience and social competences; - Participation through three modes of belonging: Engagement, Imagination, Alignment.
(Tippett et al., 2005)	Active participation	Key factors supporting social learning: - providing sufficient time, - involving stakeholders early and - careful attention to process management.
(Collins & Ison, 2009)	Social Learning for Adaptation	- Learning understood as active doing rather than mere participation; - Learning in a collective form.
(Reed et al., 2010)	Learning through social interaction	Social learning involves: - demonstrating that individuals have undergone a shift in their understanding; - embedding this change within larger social units or communities of practice; - occurring through social interactions and processes between actors within a given social network.
(Kristjanson et al., 2014)	Gathering evidence on the impact of social learning	Six-step process: - Taking stock; - Assessing options; - Getting it right; - Gathering evidence; - Analysing the evidence; - Dissemination.

According to educational theorist Etienne Wenger, social learning is "an interplay between social competence and personal experience. It is a dynamic, two-way relationship between people and the social learning system in which they participate. It combines personal transformation with the evolution of social structures" (Wenger, 2000, p. 227). Wenger distances himself from Bandura's theory by placing human experience – rooted in biological, linguistic, cultural and historical development – at the centre. According to the presented approach, the model's foundation is social participation, predicated on three distinct forms of belonging: engagement, imagination, and alignment. Engagement involves joint participation, how people engage in common activities and shape shared experiences. Imagination encompasses developing a sense of self, community, and a comprehensive understanding of the world, which is essential for individuals to make informed decisions and navigate various situations; this involves exploring available options and determining one's place in a given context. Alignment ensures that one's actions are compatible with other processes to achieve effectiveness beyond one's personal involvement.

On the other hand, Joanne Tippett and her team note that engaging in a participatory process, aimed at informed decision-making, particularly pro-environmental, does not guarantee the automatic occurrence of social learning. They highlight numerous obstacles to successful learning. "Traditional institutional structures and cultures unfavourable to participatory processes can make it difficult to gain sufficient resources and support for an ongoing process of interaction with stakeholders. Hierarchical decision-making processes can impede interaction and communication between different sectors and levels of scale. A technocratic culture, in which experts are not familiar with talking to different stakeholders in terms that they can relate to, can impede constructive communication with different stakeholders" (Tippett et al., 2005, p. 297). Therefore, effective community engagement to enhance social capital necessitates precise coordination of the process, grounded in understanding the interconnections within the social surroundings and implementing tailored procedures for management (see Figure 2).

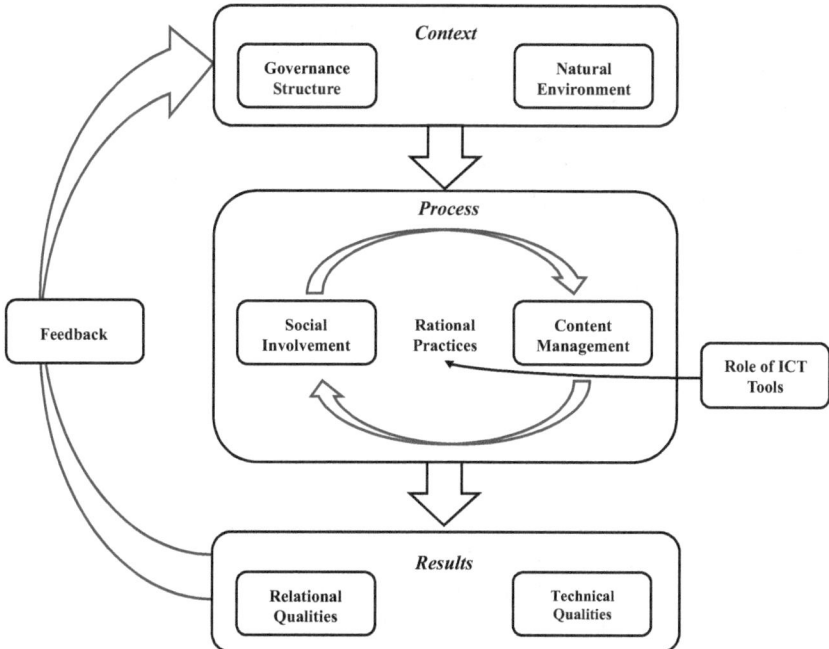

Figure 2. Conceptual framework for social learning according J. Tippett, B. Searle, C. Pahl-Wostl and Y. Rees, (Tippett et al., 2005)

Kevin Collins and Ray Ison acknowledged the necessity to revise views of social learning owing to the requirement for adaptation, which is comprehended as a co-evolutionary course triggered by climate change. They propose a social learning concept (see Figure 3) in which information, consultation, and participation may be necessary but insufficient to enhance complex situations. Therefore, advocate analysing it from a systems thinking perspective. The authors define social learning as:
- "The convergence of goals (more usefully expressed as agreement about purpose), criteria and knowledge leading to awareness of mutual expectations and the building of relational capital (a dynamic form of capital that integrates the other forms, viz artificial, natural, social and human).
- The process of co-creation of knowledge, which provides insight into the causes of, and the means required to transform, a situation. Social learning is thus an integral part of the make-up of concerted action.
- The change of behaviours and actions resulting from understanding something through action ('knowing') and leading to concerted action.
- Arising from these, social learning is thus an emergent property of the process to transform a situation" (Collins & Ison, 2009, pp. 364–365).

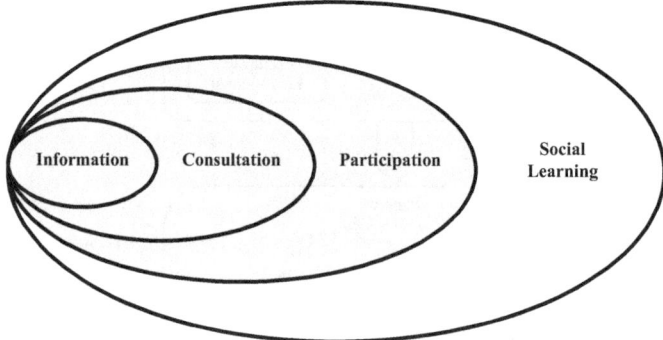

Figure 3. A conceptual framework of social learning according K. Collins and R. Ison, (Collins & Ison, 2009, p. 369)

In turn, Mark S. Reed and his team have proposed a concept based on their research into the circumstances surrounding social learning and the associated processes. They note that social learning occurs when certain conditions are met, specifically the process must: "(1) Demonstrate that a change in understanding has taken place in the individuals involved. This may be at a surface level, e. g., via recall of new information, or deeper levels, e.g., demonstrated by change in attitudes, world views or epistemological beliefs; (2) Go beyond the individual to become situated within wider social units or communities of practice within society; and (3) Occur through social interactions and processes between actors within a social network, either through direct interaction, e.g., conversation, or through other media, e.g., mass media, telephone, or Web 2.0 applications" (Reed et al., 2010).

In turn, Patti Kristjanson and her team, presented a concept based on a set of practical tools and guidelines to effectively utilise the social learning process in relation to available knowledge, information and implementation and monitoring tools. The process involves six successive steps, as presented in Figure 4. As a starting point, collecting foundational details allows for understanding of the transformation procedure and assessing whether social learning is a suitable strategy for the activity. In a second stage, feasible alternatives and solutions are identified in collaboration with the research users who will implement them. In step three, fieldworkers provide feedback to evaluate the effectiveness of the techniques employed. In the fourth stage, evidence is gathered to confirm the occurring changes, which are then analysed and interpreted by the concerned personnel to devise fresh actions and solutions. In the final crucial phase, the research data is stored and subsequently distributed for broader accessibility.

In addition to the natural trend of Bandura's social learning theory being developed by theoretical researchers, it is just as readily applied to explain phenomena occurring in a variety of fields. It has therefore also naturally attracted the attention of researchers investigating the adaptation of renewable energy-based solutions. Eva Heiskanen and her team examined the extent to which educational interventions that entail visits to residents' homes facilitated local decarbonisation efforts. Based on their qualitative study, the authors argue that the 'energy walks' can serve as a means of social learning through imitation, and identification, as proposed by A. Bandura. Additionally, they can enhance the capacity of community members to explore, analyse, and negotiate the challenges of green transformation (Heiskanen et al., 2017).

Based on a case study, Victoria Pellicer-Sifres and her co-authors (2018) investigated the internal and external factors that prompt social learning on energy-related matters within Som Energia, Spain's first renewable energy cooperative. The authors examined two levels of learning: first-order instrumental learning (the acquisition of knowledge) and second-order critical learning (the analysis, interpretation, evaluation and synthesis of information). They also explored how micropolitical and macropolitical factors act as triggers and shaping forces of social learning. The establishment of the energy cooperative has facilitated local residents in acquiring expertise on energy system legislation and market regulations (first-order learning), as well as enabling them to formulate collaborative approaches for local decarbonisation (second-order learning) (Pellicer-Sifres et al., 2018).

The adoption of solar-based solutions has attracted the interest of scholars worldwide. Based on a quantitative survey of 2,065 Canadian residents, Parkins et

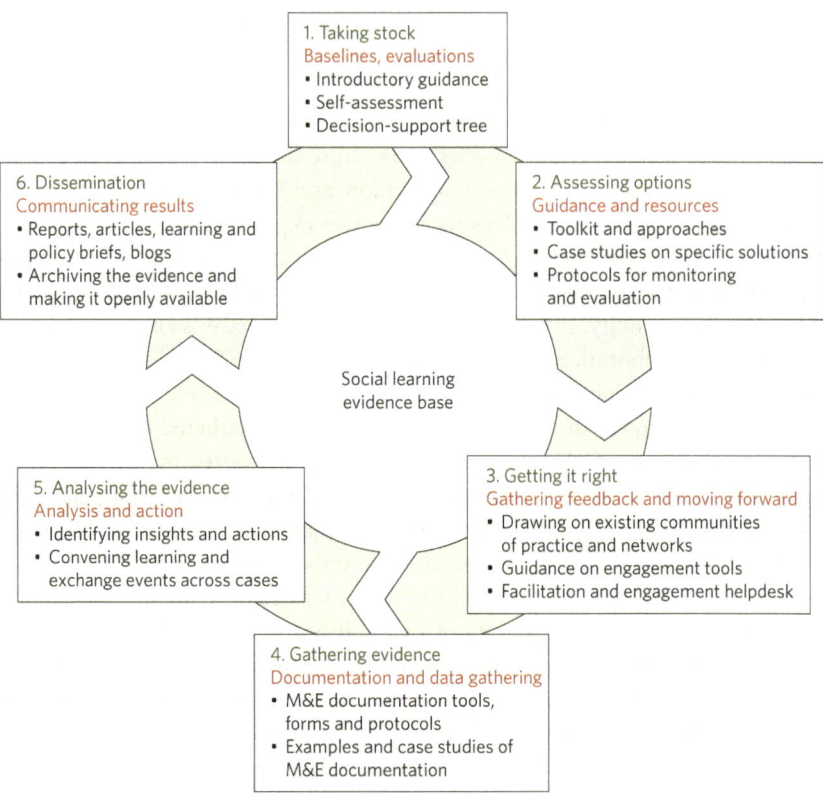

Figure 4. An evaluative framework of social learning impacts according P. Kristjanson, B. Harvey, M. Van Epp and P. K. Thornton, (Kristjanson et al., 2014, p. 7)

al. (2018) identified determinants influencing inclinations toward solar panel adoption, including age, gender, level of energy-related involvement, exposure to solar technology, marketing efforts, and educative initiatives, among others. Local policies aimed at increasing public engagement and the visibility of solar technology in communities can be more effectively planned with the support of social learning theory. Diyi Liu et al. (2023) reported from a quantitative study that active social communication on Photovoltaics (PV) installations is more effective than mere observation alone. The adoption rate of a solar thermal system is more likely to increase when influenced by social factors such as the impact of neighbours, family, and friends. Similarly, recommendations from local authorities prove more effective than passive methods such as leaflets. Additionally, the impact of social learning surpasses that of government policy (Liu et al., 2023).

Chapter 2.
Peer Effects

Each individual is part of multiple communities including family, neighbourhood, city residents, school community, co-workers, parish members, sports teams, political organisations, friends and colleagues, as well as more temporary groups like seminar or tourist trip participants, each of which can exert varying degrees of influence on their behaviour. Additionally, individuals can influence the behaviour of others, thereby altering the conditions or environment in which they operate. Technological development and globalisation have increased the influence of people who do not need to be known directly. For example, being a member of social media groups dedicated to common hobbies, interests, or political views is a clear illustration. The pandemic experience has demonstrated how quickly physical isolation can be mitigated by a network facilitated by technological advancements. Humans, as social beings, are naturally subject to social interactions within various social networks, whether desired or not. In the theory of diffusion of innovations, which will be discussed in detail in the next chapter, peer groups are referred to as a social system, defined as "a set of interrelated units that are engaged in joint problem-solving to accomplish a common goal. The members or units of a social system may be individuals, informal groups, organisations, and/or subsystems. A system has structure, defined as the patterned arrangements of the units in a system, which gives stability and regularity to individual behaviour in a system. The social and communication structure of a system facilitates or impedes the diffusion of innovations in the system" (Rogers, 1983, pp. 24, 37).

The typology of peer effects (Figure 5) presented below is based on the categorisation proposed by Weihua An (2011).

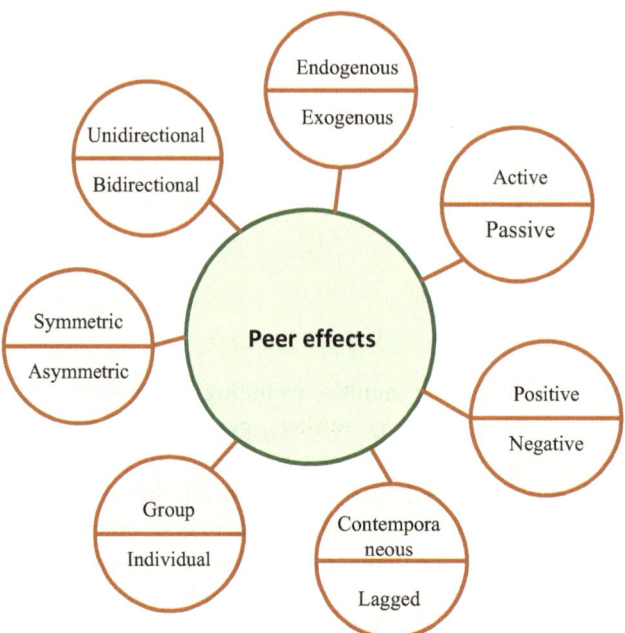

Figure 5. The typology of peer effects. Author's own elaboration based on Weihua An (2011)

Endogenous and Exogenous Peer Effects

Peer effects refer to situations in which behaviours, attitudes, decisions shape and are shaped by interactions within a peer group. This term extends beyond the social sciences and encompasses a variety of complex situations. Endogenous peer effects arise from mutual interactions and influences among individuals within a social network. In contrast to exogenous peer effects, which stem from external factors such as neighbourhood characteristics or random assignment to a group, endogenous peer effects emerge from the dynamics of relationships and ongoing social interactions. These effects may manifest in various forms, including social contagion, social reinforcement, social comparisons, and social learning, wherein individuals adjust their behaviours or attitudes in response to observed behaviours or perceived norms within their social environment.

The identification and empirical measurement of peer effects can be methodologically challenging due to the complex interdependencies among individuals in social networks and potential confounding factors. The presence of a peer effect is not obvious and often equates to a correlation between individual peer behaviours. Charles F. Manski observes that comparable group behaviour is encouraged under certain circumstances, making it difficult to clearly identify peer effects:

> "(a) *endogenous effects*, wherein the propensity of an individual to behave in some way varies with the behaviour of the group;
> (b) *exogenous (contextual) effects*, wherein the propensity of an individual to behave in some way varies with the exogenous characteristics of the group, and
> (c) *correlated effects*, wherein individuals in the same group tend to behave similarly because they have similar individual characteristics or face similar institutional environments" (Manski, 1993, pp. 532–533).

Endogenous peer effects operate through an intricate interplay of cognitive, emotional, and social processes that shape individuals' perceptions, motivations, and decision-making. Social identity theory (Tajfel & Turner, 1979) posits that individuals derive a sense of identity and belonging from membership in social groups, leading them to conform to group norms and behaviours to maintain cohesion and social acceptance. Social comparison theory (Festinger, 1954) suggests that individuals evaluate their own abilities, attitudes, and behaviours in relation to peers, thereby influencing their self-perception and behaviour accordingly. Additionally, social learning mechanisms such as imitation, emulation, and observational learning enable individuals to acquire new skills, knowledge, and behaviours from their peers, thereby facilitating the diffusion of innovations and cultural practices within social networks.

In the context of energy transformation, these types can be illustrated through the following examples. Endogenous effects, as defined by C.F. Mansky, occur when an individual's behaviour changes specifically due to the incidence of similar behaviour among the peer group to which that individual belongs. This refers to the direct influence on an individual of the behaviour or decisions made by other group members. Consider an educational intervention aimed at selected residents of an apartment block to increase their knowledge of how to save electricity. As a result of this intervention, the average monthly energy consumption of the apartment block decreases. Consequently, in the following month, the possibility of reducing energy consumption (based on overall average statements) became apparent. As a result, other residents began to adopt similar behaviours, indirectly influencing the entire community. Conversely, excessive energy consumption by a select group can increase in the average consumption for the entire block, which may reduce the willingness of other residents to save energy. However, the presence of a correlation between the behavioural change of people from both groups – those who participated in the educational intervention and those who did not – does not necessarily indicate that the behavioural change is the result of peer effects. To diagnose the presence of endogenous peer effects, researchers should understand from the outset how individuals form their reference groups.

Toshi H. Arimura, Hajime Katayama, and Mari Sakudo (2016) conducted a study on the impact of social norms among close friends on energy conservation practices, exploring endogenous effects. The authors utilised data from a household survey conducted in Japan for this study. Positive endogenous social effects were found in both summer (0.185) and winter (0.556), but they were not statistically significant. Therefore, the authors claim there is insufficient evidence there is insufficient evidence that people are influenced by their peers in their decision to save energy, the authors claim. However, the results may have been affected by the survey instruments used in this study, which assessed social norms through self-report. In fact, energy-saving behaviour may not be readily observable by acquaintances. Therefore, individuals' self-reported perceptions of their friends' home temperature preferences may reflect not only their friends' actual behaviour but also their own expectations of how people should behave. The authors suggest that omitted variables, which influence both an individual's behaviour and others' perceptions of their behaviour, may actually be responsible for the endogenous social effects identified (Arimura et al., 2016).

Exogenous effects occur when an individual's behaviour depends on the characteristics of the peer group, i.e. the reference context. Different frames of reference can be induced by external factors such as social norms, cultural influences, or institutional policies, which in turn influence the behaviour and outcomes of individuals within peer groups. In contrast to endogenous peer effects, which arise from interactions within the peer group itself, exogenous peer effects stem from external factors relative to the immediate social context but still impact individuals within that context. Social norms play a significant role as individuals often conform to social expectations and standards within their peer groups. Cultural influences, including values, beliefs, and traditions, shape individuals' attitudes and behaviours through socialisation processes. Additionally, institutional policies such as educational reforms or initiatives addressing pollution control and environmental protection can influence peer groups by shaping the broader social environment in which individuals interact. Similarly, economic policies that affect income distribution, access to resources, subsidies, or grants for technological innovations, can shape peer effects in terms of economic behaviours and outcomes. Social policies aimed at reducing inequalities or addressing social challenges such as discrimination or poverty, including energy poverty, can also impact peer groups by altering the social environment in which individuals interact (Primc & Slabe-Erker, 2020).

It is crucial to exclude subjective evaluations in assessing the presence of exogenous effects, unless they are clearly marked as such. It is important to note that these effects are not the result of peer influence but rather the result of the specific conditions in which individuals find themselves. For instance, if a group of individuals reside in a housing estate constructed by the same developer using a similar building design, the residents may be less likely to install photovoltaic panels if the building design makes it challenging to do so. Conversely, if a building's design is well-suited for the installation of photovoltaic panels, residents may be more inclined to install them compared to residents in other areas. Thus, although there are strong correlations between the observed behaviours, they are not necessarily the result of a peer effect. When examining a case, it is important to consider other factors that may influence the observed correlations and be the actual cause of their occurrence. Identifying and disentangling exogenous effects from endogenous ones can pose a methodological challenge, requiring advanced research designs and data analysis techniques that account for the complexity of social interactions and the diverse contexts in which they occur.

A third type of circumstance that makes it difficult to observe the occurrence of a peer effect, as pointed out by C.F. Mansky, are correlated effects, which result from the similarity within the peer group itself. The selection of a residential location by a specific group of people may be influenced by certain features of the area, such as a suburban neighbourhood with abundant greenery, which attracts individuals with a similar appreciation for nature. Consequently, environmental concern is a direct outcome of the characteristics of these individuals, rather than a peer effect among a community residing in the same locality.

Research on peer effects among Xiamen University students regarding the adoption of environmentally friendly behaviours related to energy conservation, focusing on endogenous, exogenous, and correlated effects, was conducted by Boqiang Lin and Huanyu Jia (2023). In their conducted research, the authors tested the hypothesis that peer incentives constitute an environment for mutual learning among students facilitating the development of an ecologically sustainable lifestyle. They demonstrated that peers' green behaviours have a positive impact on the energy-saving behaviours of individual students. Specifically, with an increase in energy-saving behaviours among peers by one standard deviation, similar behaviours of the respondents improved by 0.132 standard deviations. This finding provides valuable insight, particularly for individuals planning to implement actions or educational campaigns. The authors also argue that individuals in different classes, with varying levels of depression, environmental concerns, and peer composition, are subject to different levels of peer influence, exogenous effects (Lin & Jia, 2023).

Active and Passive Peer Effects

Another type of peer effects are those that engage individuals' attention to varying degrees and conditions. According to Weihua An (2011, p. 515), "active peer effects come from connections that a person can explicitly recognize while passive peer effects come from peers that a subject does not have an explicit tie with. Friends' effects are examples of the former. Transmission of infectious diseases or market competition can serve as examples of the latter". Active peer effects, according to this definition, arise from conscious actions and may thus e demonstrate the ability to induce motivation. Within peer environments, individuals often feel more motivated to act because they see their peers also making efforts towards achieving their goals. For instance, participation in health-promoting activities may stimulate motivation to take additional actions to combat pollution in one's living area, driven by the desire to achieve common objectives.

Active peer effects also foster mutual support. When people with similar interests or goals come together, they can share knowledge, experiences, and strategies which increases the likelihood of making difficult decisions. For example, through involvement in regular community meetings, residents may collectively decide to invest in a local micro-power plant and mutually motivate each other to reduce the building' energy consumption. Active peer effects can thus support pro-social activities and enhance engagement. When individuals collaborate on common projects or social initiatives, they can achieve significantly more than they would independently.

Tyler J. VanderWeele and Weihua An (2012) extend this framing by defining passive peer effects as those that occur naturally and active peer effects as those that are additionally supported by persuasive actions. The authors explore the question of "whether to train or incentivize the change agents to actively advocate for the proposed diffusion (namely, to utilize active peer effects). For example, in a smoking prevention program, would training of the treated subjects and providing monetary incentives for them to actively advocate their peers not to smoke make the intervention more effective? This consideration is consequential for the selection of central subjects. For example, to accelerate active peer effects using training, it may be that indegree (the number of received ties) is the more important characteristics, while if the purpose is to accelerate passive peer effects (namely, to utilize peer effects in their natural state, without training or extra incentives), outdegree (the number of outgoing ties) may be more important" (VanderWeele & An, 2013, pp. 359–360).

Passive peer effects refer to the subtle influence individuals exert on their environment through their behaviour, attitudes, and personality. These effects arise from daily interactions within peer groups. Even without direct interaction or manipulation, individuals can influence others through their presence, demeanour, or actions.

Varun Rai and Scott A Robinson later defined active and passive peer effects concerning the adoption of energy technologies, with peers identified as neighbours living within a five to ten block area. According to them "Passive peer effects refer solely to the attitudinal and behavioral stimulus that seeing PV systems in the neighborhood induces. It excludes the effect of contact with other PV owners" (Rai & Robinson, 2013, p. 3), while active peer effects refer to "influence that accrues through peer-to-peer communication through contact with neighbors" (Rai & Robinson, 2013, p. 7). The authors demonstrate that passive peer effects occur when individuals observe PV installation of their neighbours. This process builds trust in PV technology and increases the motivation to install it on one's own property. Active peer effects, on the other hand, operate through direct, interpersonal contacts between neighbours, resulting in a 4.6-month reduction in the adoption time of a given technological solution. In contrast, when both passive and active peer effects are absent, there is an increase in adoption time by 6.67 months (Rai & Robinson, 2013).

Research by Alvar Palm (2017) suggests that active peer effects play a more significant role than passive effects. However, the authors of the qualitative research note that this finding may be influenced by the high awareness of the respondents, who already possessed substantial knowledge about PV systems before observing their neighbours' installations. Additionally, it is important to consider that the research methodology used might have influenced the study's findings. The interviews conducted relied on the subjective opinions of the participants, potentially emphasising active peer effects more than passive effects. Passive effects may indeed play a significant role, as exposure to views of operational photovoltaic systems could prompt attempts at direct contact with their owners, potentially reducing adoption time for new technologies. The adoption of renewable energy sources by neighbours can be seen as a positive example. However, receiving feedback in a face-to-face conversation that a neighbour is satisfied with such technology is more likely to encourage others to make a similar choice. This effect could be particularly pronounced for technology solutions requiring substantial financial investment (Palm, 2017).

Research also indicates that the decision to adopt photovoltaics is influenced by peer perception. According to survey research conducted by Fabiana Scheller, Sören Graupner, James Edwards, Jann Weinand, and Thomas Bruckner, adoption decisions occur more quickly when peers who have installed a photovoltaic system are perceived as credible individuals. In such scenarios, efforts to initiate peer interactions regarding photovoltaic systems are also more frequent. The authors also suggest that as the size of the decision-maker's peer group grows, and as the decision-making process regarding the adoption of technological innovation progresses, the perceived credibility of peers as a source of recommendation for photovoltaic systems also increases. Furthermore, decision-makers often make critical decisions regarding energy issues without realising that they have been influenced by peers or stakeholders. Thus, at various stages of crucial decision-making, individuals may be influenced to varying degrees by others without necessarily being aware of it (Scheller et al., 2022).

Positive and Negative / Desirable and Undesirable Peer Effects

Another type of peer effect focuses on its direction. According to Daniel Noll, Colleen Dawes and Varun Rai, "positive peer effects in the context of PV can be understood to influence a consumer decision by providing information that (a) demonstrates the relative advantages of solar, (b) proves the compatibility of solar with a consumer's existing beliefs and habits, (c) reduces perceptions of solar technology's complexity, and (d) shows the results of the trials of other installations. Positive peer effects for PV can also increase the likelihood of adoption and decrease the length of decision time" (Noll et al., 2014, p. 3). Building on this definition of peer effects in renewable energy technology, negative or undesirable peer effects occur when social interactions diminish the likelihood of adopting technological solutions that promote energy system transformation, prolong the decision-making time for such solutions, or result in dysfunctional outcomes.

Research on positive and negative peer effects was conducted by Daniel A. Brent, Joseph H. Cook and Allison Lassiter (2022). They focused on programmes in the United States aimed at increasing the adoption of green stormwater infrastructure to capture rainwater. One such program was the voluntary RainWise programme, which provided grants for rain gardens and water harvesting cisterns. The authors highlight the presence of positive peer effects among residents in neighbourhoods where more individuals had installed rainwater harvesting devices. These individuals, as part of the program's requirements, also displayed posters provided by the utility company (Brent et al., 2022). Bryan Bollinger and Kenneth Gillingham (2012) also highlight the role of positive neighbourhood peer effects in green transformation through the placement of demonstration plaques. Conversely, the negative peer effects observed by the authors of the aforementioned study involved a group of residents of more expensive houses who were required to use private green stormwater infrastructure. These installations were reported less functional and attractive compared to RainWise (Brent et al., 2022). Eric O'Shaughnessy and co-authors (O'Shaughnessy et al., 2020; Cook et al., 2021) observe that sharing negative experiences of using an environmental installation, as well as about the installation process itself, can lead to negative peer effects or at least counteract positive peer effects.

Contemporaneous and Lagged Peer Effects

Another type of peer effects are those that focus on the temporal aspect of their impact. Situations where we can observe the effect of a social interaction immediately are referred to as contemporaneous peer effects. An example of this would be the abandonment of using plastic bags for group shopping in response to the behaviour of fellow shoppers who use reusable cloth bags. Another example could be employees demonstrating increased commitment when working with more capable colleagues (Kato & Shu, 2009) or showing greater willingness to collaborate with another branch when their teammates do so (Aschhoff & Grimpe, 2014).

Tat Y. Chan, Jia Li and Lamar Pierce (2014a) analysed workplace interactions to investigate how peer-based learning can enhance salesperson skills. The study focused on contemporaneous peer effects by comparing worker productivity. It considered the selection of coworkers, including stars of the week and less skilled workers, in relation to shift assignment policies. They proposed a new approach to estimate this learning effect by considering the accumulation of contemporaneous peer effect as 'human capital' that will influence future sales. According to the authors' research, working with high-skill peers has a significant impact on the long-term productivity growth of new salespeople (Chan et al., 2014a).

Situations where the effects of social interactions are observed with a delay are term lagged peer effects. These effects highlight the longitudinal nature of social influence, where the impact of peer interactions may not be immediately apparent but can manifest themselves over long term. Examples include avoiding plastic bags for shopping in response to a school educational campaign, adopting new technologies in agriculture as a result of previous decisions by farmers in the same village (Larson et al., 2016), or delaying the installation of photovoltaic systems based on guidance from previous users (O'Shaughnessy et al., 2023). The lag period's length is not defined from a top-down approach; rather, it is defined individually by researchers based on the study's methodology and context. For instance, Michael D. Frakes and Melissa F. Wasserman (2018) adopted a two-year period to examine the effect of peer influence on patent office examiners' award decisions, whereas Tom Fangyun Tan and Serguei Netessine (2019) used a one-week period to study how peer influence affects restaurant associates' sales performance.

Group and Individual Peer Effects

Peer effects can be based on individual social connections between two people, known as dyads (Moreland, 2010), or on interactions within a group. Individual peer effects refer to the influence that individual members of a peer group have on each other. We can talk about situations where peers such as spouses, friends, or close colleagues -for example a director and their assistant or a chairman and their advisor – can influence each other. Even when these individuals are part of a larger peer group, individual effects pertain to situations where the behavioural changes or decisions made as a result of the social interaction taking place are independent of the behaviour or decisions of other community members (Rogers, 1983) – for example, the decision to install a rainwater recovery system made by the owner of an individual detached single-family house.

Group peer effects, on the other hand, occur when behavioural change or decision-making happens in relation to a group of individuals and is often based not only on interaction among group members but also on their cooperation. It is not uncommon for similar behaviour or decisions made to require adaptation to the behaviour or decisions of the group. In social groups, peer effects can be observed in various situations, such as educational attainment within a class, in decision-making by a group of experts providing a joint opinion, or among colleagues working the same shift, as mentioned earlier.

The differentiation between individual and group effects is significant not only due to the type of interactions involved but also because individuals may exhibit varying behaviours in diverse social situations. When accompanied by a friend, individuals may discard a leaflet found on their windscreen onto the pavement. However, when in the company of superiors from work, they are less likely to. Instead, they are more inclined to throw the leaflet into a bin or keep it in the car. Jianxing Wu and colleagues conducted research on the behaviour of tourists in three distinct social contexts: interactions with family, friends, and strangers. The results of their study indicate that individuals who behave in an environmentally friendly way in the presence of parents or spouses may alter their behaviour in the company of acquaintances or strangers, potentially following others in engaging in environmentally unfriendly actions, such as leaving litter in the forest (Wu et al., 2023).

Shruti Gupta and Denise T. Ogden (2009) conducted a study to investigate consumer attitudes towards environmentally friendly products or services within the context of peer effects. The study aimed to determine the reasons why consumers choose to purchase or refrain from purchasing such products or services based on their attitudes towards the environment. The authors propose that the choice to purchase environmentally friendly products is influenced by the identification with a reference group, such as green consumers or celebrities.

This leads to a greater tendency to act for the collective benefit of the group rather than individual benefit. Green buyers were more likely than non-green buyers to make cooperative decisions when purchasing environmentally friendly products expecting other consumers to behave similarly. They were therefore more cooperative than consumers who identified themselves to be non-ecological (Gupta & Ogden, 2009).

Unidirectional and Bidirectional Peer Effects

"Unidirectional peer effects occur when peer effects flow only one way, from one subject to another, but not the other way around. Bidirectional peer effects happen that peers influence each other and the peer effects flow reciprocally" (An, 2011, p. 515). Unidirectional peer effects can stem from various mechanisms, an individual's stronger personality, knowledge, or social position within the group. Individuals with enhanced communication or leadership skills may exert greater influence on their peers, shaping their attitudes and behaviours. Additionally, individuals with higher social prestige or knowledge may be more credible and persuasive to their peers, enabling them to exert stronger influence. When considering the adoption of a particular energy system, research often focuses on unidirectional peer effects which involve examining situations where the owner of the system influences the decisions of potential users.

Thus, for example, Chen Qing and his team studied the influence of relatives and friends on biogas adoption decisions in 540 rural households in China. The peer groups were categorised into two subgroups: those who visited the household during Chinese New Year, which were classified as representing strong ties with the household member under study, and relatives and friends who did not visit the household during Chinese New Year, which were classified as representing weak ties with the household member under study. Qing et al. (2022) identified three main findings from their study. Firstly, they observed that the decision of rural households to adopt biogas is significantly influenced by the adoption of biogas by their relatives and friends. Secondly, it appears that the decision to adopt biogas on farms was more influenced by relatives and friends who have strong ties to the surveyed households, compared to peers who have weak ties. Furthermore, a stronger peer effect was observed among farmers with lower levels of education and those who are farther away from the market (Qing et al., 2022).

Symmetric and Asymmetric Peer Effects

When exploring bilateral peer effects, it is also worth comparing the size of the effects that occur in each direction. A symmetrical effect occurs when a person's influence on a peer is of the same magnitude as the peer's influence on that person. Conversely, the magnitudes of the individual effects are different, the effect is considered asymmetric.

Tat Y. Chan and colleagues (2014b) conducted an analysis to examine the potentially asymmetric nature of peer effects among employees with different skills working in the same teams within a team-based compensation system. The system is designed to support employees' motivation to help each other while also encouraging competition. According to a study conducted among cosmetics salespeople in a department store, it was found that employees with higher abilities have a more significant impact on increasing sales productivity compared to employees with lower abilities have on decreasing sales productivity.

While there may be different perspectives on peer effects, it is widely acknowledged that they play a significant role in influencing decisions to adopt energy transition measures. In contemporary society, it can be challenging to escape the influence of factors such as family, friends, media, and sophisticated marketing strategies that encourage exploring new opportunities to enhance the quality of life and well-being of individuals and communities. Therefore, in the following chapter, I will delve into the mechanisms that contribute to the formation of peer effects.

Chapter 3.
Mechanisms of Peer Effects Formation: Communication Channels

Green transformation is a process leading to "transition to a climate-neutral, sustainable, non-toxic, resource-efficient, renewable energy-based, resilient and competitive circular economy in a just, equitable and inclusive way, and to protect, restore and improve the state of the environment" (European Union, 2022, p. C 243/35). This transformation involves adjusting attitudes, preferences, behaviours, and values towards environmental quality, making it a complex and multidimensional process. It spans from simple actions like using reusable bags when shopping, to more deliberate decisions such as adopting renewable energy sources in households, and extends to highly complex endeavours involving industry, corporations or the implementation of national and international policies. As this monograph aims to explore how social learning can either facilitate or hinder the adoption of renewable energy solutions among individual households, this chapter will focus on understanding how peer effects constitute a crucial factor influencing the energy transition process.

These mechanisms will be analysed through the lens of diffusion of innovations theory, for the decision to adopt or install a renewable energy system exemplifies such a process – specifically the process of adoption of a technological innovation.

Diffusion of Innovation Theory

The theory of diffusion of innovations (DoI) was popularised by Everett M. Rogers, an American sociologist who first presented it in his 1962 book, "Diffusion of Innovations". Rogers aimed to address how to accelerate the gradual process of adoption and adaptation of new ideas by society. Rogers distinguishes four key elements of the diffusion of innovation (or technology or technological innovation, terms Roger uses interchangeably), shown in Figure 6:
1. The first component is innovation, which may be defined as an idea that is perceived by an individual as novel or new.

2. The second component is the transfer of innovation between individuals. Diffusion is defined as the process whereby an innovation disseminates from one person to another.
3. The third component is the social system within which the innovation is disseminated. In this conceptualisation, a social system is understood as a population of individuals whose functional differentiation enables collective problem-solving processes. A diffusion study may analyse a social system comprising all farmers in a given area, all doctors in the local community, or members of an indigenous tribe.
4. The final component is the period during which the innovation is adopted by individuals within the social system. Adoption may be defined as a decision to utilise a particular innovation to the fullest extent possible. The adoption of an innovation does not occur simultaneously among all individuals. Innovativeness may be defined as a measure of the extent to which an individual is more likely than their social group to adopt new ideas at an early stage (Rogers, 1962, pp. 1–2).

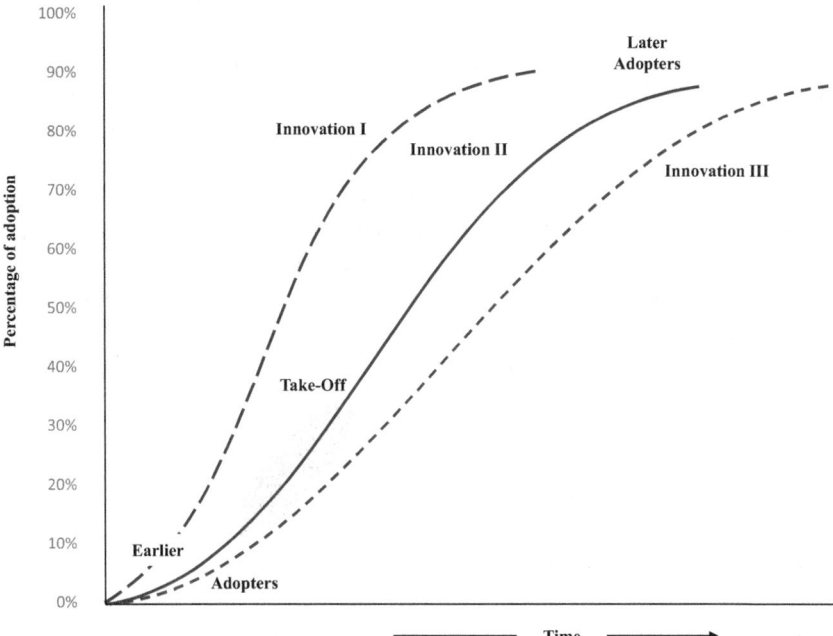

Figure 6. Diffusion is the process by which (1) an innovation (2) is communicated through certain channels (3) over time (4) among the members of a social system. Source: (Rogers, 1983, p. 11)

Each of the four key elements of innovation diffusion has specific characteristics. Thus, the first key element – innovation, is characterised by five attributes. The first attribute is 'Relative advantage', which describes the degree to which a new technology is perceived to be better than the technology already in use. There is no predefined unit for this measure, as it can be interpreted differently by different users. It may refer to a solution that is more favourable in terms of factors such as economy, convenience, or satisfaction.

The second characteristic of innovation is 'Compatibility', which determines how well the new technology meets the current needs, expectations, or values of potential users. If compatibility is high, the innovation is more likely to be adopted; otherwise, the probability is lower or more uncertain. Using the example cited earlier, if the construction of a residential building requires no additional work and the location's climate provides ample available solar energy, the building owner can be more easily and quickly encouraged to install photovoltaic panels.

According to Rogers, 'Complexity' is the third attribute of innovation, referring to the level of difficulty that potential users perceive or understand about a technology. If people perceive a technology as complex to install and operate, the time required for its acceptance will be longer compared to technologies that are perceived as easier to understand and use.

The fourth attribute of an innovation is 'Trialability', which refers to the possibility of testing it over time. According to a questionnaire study conducted by Nirmal M. Menon and I. Sujatha (2020) among solar panel users from Thrissur, India, those who had the opportunity to attend pre-purchase events demonstrating the actual operation of the system made purchase decisions more quickly than those who did not attend such events (Menon & Sujatha, 2021).

Finally, the fifth attribute of an innovation is 'Observability', which refers to the degree of visibility of the innovation. The ability to observe a technological innovation, its performance and its effects more frequently increases the likelihood of adoption of that innovation (Rogers, 1983, pp. 15–16). Factors such as relative advantage, compatibility, complexity, trialability and observability influence the rate and extent of adoption, shaping the diffusion curve over time.

The following sections of this chapter will detail the communication channels that constitute the second key element of the innovation diffusion process. These channels serve as examples of mechanisms for creating peer effects.

According to DoI theory, the decision-making process for adopting an innovation takes place over time and consists of five steps (Figure 7). The initial stage involves acquiring 'Knowledge' about innovation. This can be intentional, such as attending a training course, or unintentional, such as having a casual conversation with a fellow passenger on a train. Information is gathered about the technology, its workings, or its benefits. During step two, in the 'Persuasion' stage, individuals develop an attitude towards the innovation, which may be either positive or negative. The next step is the 'Decision' stage, when the person takes actions that lead to the adoption or rejection of the technology. Step four is the 'Implementation' stage, when the innovation is put into use. The fifth and final stage is 'Confirmation'. During this stage, the adopter either confirms that the decision to innovate was the right one and continues to use the adopted technology, or abandons it if the adopted technology has not met their expectations to a significant degree (Rogers, 1983, p. 20).

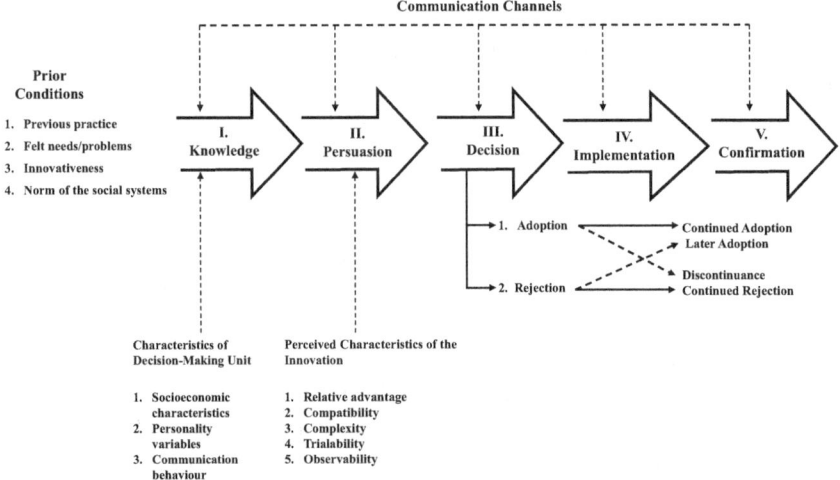

Figure 7. The Innovation-Decision Process. Source: (Rogers, 1983, p. 165)

The fourth key element of innovation diffusion, as mentioned earlier, involves the members of a social system. Since it is very common for adopters of an innovation to have peer relationships with each other, Rogers introduced five categories of adopters to help distinguish their status and connections: (1) innovators, (2) early adopters, (3) early majority, (4) late majority, and (5) laggards (Figure 8). The percentages of the different categories of adopters shown in Figure 8 follow a Gaussian curve distribution. The rate of adoption is influenced by the time it takes for the members of the social system to adopt the technological innovation. According to Rogers, the adoption rate refers to the duration it takes for a certain percentage of adopters to accept an innovation. According to Figure 6, during the initial stage of the diffusion process, a small group of individuals, commonly referred to as innovators, are the first group to adopt the technology. The presence of such characteristics can be observed in the initial segment of the S-curve. Innovators are individuals who demonstrate an entrepreneurial spirit and a willingness to explore new ideas, even if it involves taking risks and requires courage. In the diffusion process, innovators can play a crucial role as they tend to be well-travelled and open to sharing new ideas across various peer groups.

Over time, innovations tend to spread through various channels, leading to an increasing number of adopters and a climbing trajectory of adoption rates. A further category of innovation adopters that arises are early adopters. According to Rogers, the group is typically local but exhibits a notable level of leadership within a peer group or local community. They are eager to share their knowledge of the innovation with other potential adopters. Through social interaction, early adopters influence the early majority, a group of adopters who adopt innovations just ahead of the average peer group member. Although the early majority is relatively open to technological innovations, they tend to follow others rather than being leaders.

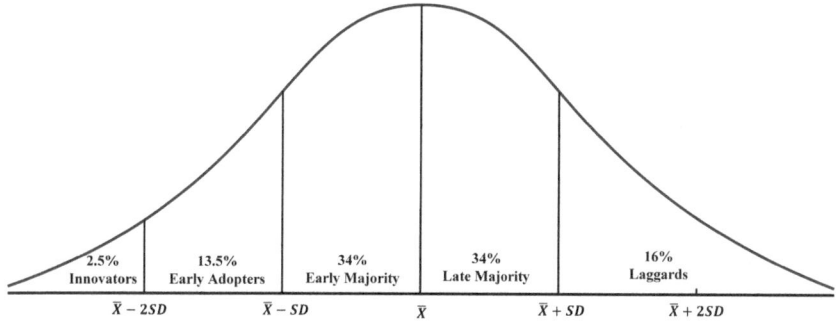

Figure 8. Categories of adopters. Source: (Rogers, 1983, p. 247)

In the next stage, the curve begins to flatten as gradual saturation occurs and fewer potential adopters remain. Among these are the late majority, who tend to adopt a technological innovations not on their own initiative but due to necessity – most often economic necessity, increasing peer pressure, or because most of their peers are already using the technological innovation in question. The final group, called laggards, are individuals who may require additional attention and are more challenging to reach and persuade to adopt the innovation. Often, they are poorly socially networked and have limited interactions with their peers, sometimes even isolating themselves. The slope of the diffusion curve is influenced by the rate of adoption of the innovation. A more rapid rate of adoption tends to result in a steeper curve, whereas a slower rate tends to result in a flatter curve (Rogers, 1983).

Despite potential benefits, the implementation of innovations often encounters barriers that impede their absorption and dissemination. Resistance to change, reluctance to take risks, inertia, and uncertainty about the outcomes of innovation can foster scepticism or complacency among stakeholders, hindering adoption. Moreover, factors such as lack of awareness, knowledge gaps, resource constraints, and compatibility issues with existing systems or practices can constitute significant barriers to innovation adoption, especially in complex or entrenched industries. Additionally, cultural norms, institutional barriers, and regulatory constraints may inhibit the adoption of innovations that challenge dominant norms or structures. By understanding these factors, barriers, and consequences of innovation implementation, stakeholders can harness the transformative potential of innovation to drive progress and prosperity. From improving quality of life to addressing global challenges, innovation implementation offers unlimited opportunities for positive change and social development. Embracing innovation requires courage, vision, and collaboration to create a future that promotes social inclusion, sustainability, and resilience in the face of unprecedented challenges, including the contemporary threats posed by climate and energy crises.

Observational Learning

The formation of peer effects as part of the information diffusion process does not necessarily imply the need to build direct peer interpersonal relationships. Observational learning occurs when new behaviour results from imitating behaviour observed in peers. Thus, in relation to energy transition, learning, including the acquisition of knowledge about technological innovation, occurs through the observation of peer solutions. In line with Bandura's (1977) social learning theory discussed in the first chapter, observational learning operates through cognitive processes such as attention, retention, reproduction, and motivation. These processes enable potential users of alternative energy solutions to acquire and internalise information from the environment in which they operate. Attention involves actively focusing on the behaviour of models, such as noticing renewable energy installations functioning in their surroundings. Retention entails encoding and storing observed information in memory, as the decision to adopt alternative energy installations is not made upon a single observation. Reproduction involves imitating behaviour, while motivation determines the likelihood of engaging in behaviour based on anticipated rewards or consequences, often economic in this context.

By observing the consequences of decisions made by other users of energy installations, potential adopters learn which decisions may be perceived as favourable or undesirable, shaping their subsequent choices and actions. Furthermore, observational learning enables individuals to learn from the mistakes and successes of others, helping them avoid pitfalls and adopt effective strategies to achieve their goals. It facilitates the dissemination of innovations, technologies, thereby driving social progress and adaptation to changing environments. In the digital age, media platforms serve as powerful channels for observational learning, influencing the attitudes, beliefs, and behaviours of individuals by exposing them to diverse media content and role models. Television, film, social media, and online platforms provide numerous opportunities for individuals to observe and emulate characters, celebrities, influential individuals, and peers, both through incidental exposure and deliberate product placement (Cook et al., 2023).

One of the more commonly cited studies attempting to estimate the size of the peer effect based on observational learning is the study by Bryan Bollinger and Kenneth Gillingham (2012). The authors proposed aggregating a complete PV installation with the postcode of each site. The study was conducted for installations established between January 2001 and December 2011 in California, USA. The authors demonstrate that each additional complete installation increases the probability of another installation in the same postcode area by 0.78 percentage points. In addition, the authors suggest that the size of the installation also matters for the adoption rate. According to their hypothesis, larger installations are more visible, and seeing them prompts neighbours to more quickly decide to replicate the technology in their own household (Bollinger & Gillingham, 2012).

Spatial studies of the diffusion of energy technological innovations are also conducted in Europe. For instance, Laura-Lucia Richter (2013) analysed the peer effect through observational learning using data on the number of solar installations in the United Kingdom (UK), focusing on an area in Scotland with an average installation density rate of 0.7 installations per 1,000 households. The author estimated that the next photovoltaic panel visible in the same location, increases the installation rate three months later by one per cent, which translates into a 0.5 percentage point increase in the number of new installations in the area. Research has also indicated that social learning through observation varies across different months and tends to decrease over time. These effects are also stronger at the level of local communities compared to the level of local authorities in geographic terms. An intriguing finding from the research highlighted by the author was that moderately affluent neighbourhoods exhibited a lower degree of innovation diffusion, indicating that social learning played a less significant role in those areas. In contrast, a stronger learning effect was observed in neighbourhoods with moderately higher education levels compared to those with lower education levels (Richter, 2013).

Nicholas Nixon Opiyo (2019) attempted to estimate the timeframes involved in the diffusion of innovative solar technologies. He investigated this issue in a location where solar PV installation is considered a lower-risk solution due to favourable climatic conditions. The author applied data obtained from a quantitative survey of 192 household residents from Kenya (villages of Gendia, Kanam, and Pala) in an agent-based model. Opiyo distinguished two groups of participants among the respondents, 18 were classified as early adopters and the remaining 174 as imitators (observers). The simulation carried out allowed to verify the hypothesis that the visibility of solar systems installed stimulates further installations in the neighbourhood. This is because it is not only the installation itself that is visible, but also due to the effects of the installation in the form of a better lighting of the farm area at night which gives the sense of security of the residents. Moreover, early adopters are perceived to have achieved a higher social status through the adoption of innovations, so the desire to emulate is also dictated by the motivation to achieve a similar status (Opiyo, 2019).

In contrast, using Dutch spatial data, Jianhua Zhang, Dimitris Ballas and Xiaolong Li (2023) suggest that an increasing number of PV installations in a neighbourhood contributes to increasing of the number of similar installations in close neighbourhoods. The visibility of the solar panels contributes not only to the further acceptance of the technological innovation, but also to the strengthening of the relationships with the peers. The authors similarly estimated an innovation adoption rate similar to Bollinger and Gillingham (2012). Using data from 13,205 Dutch neighbourhoods, they estimated its rate at 0.41 (Zhang et al., 2023).

Word-of-Mouth Effect

When considering the mechanisms for the formation of direct peer effects, it is worth starting with those derived from communication interactions. Communication channels, the second key element in the innovation diffusion process, are the means through which information about innovations is transmitted to potential adopters. There is no doubt that nowadays the spectrum of available communication channels is wide, ranging from the informal ones such as a simple social conversation or a discussion between two or more people, to the more formal ones, such as trainings, workshops or instructions. The latter will be presented in the following sections. Due to the variety of available media, communication channels play different roles both in the process of acquiring knowledge about a technological innovation and in the decision-making process to adopt it. This section delves into the multifaceted dimensions of word-of-mouth marketing, examining its mechanisms, influence on the behaviours of potential users of renewable energy-based systems, and its significance in contemporary marketing strategies. The power of word-of-mouth marketing lies in its organic nature, driven by individuals sharing their experiences, opinions, and recommendations with others.

Word-of-mouth marketing, as mentioned above, can be defined as the transmission of information between two or more people through face-to-face communication, telephone conversations, or casual or purposeful social gatherings. Word-of-mouth marketing can also occur through media, yet its character remains unchanged. It encompasses conversations held on informal online platforms such as discussion forums, review websites, social media, or instant messengers. The transmitted information may include recommendations, endorsements, reviews, and discussions about products, services related to their installation or operation, brands, or experiences. In contrast to traditional advertising, which often relies on paid media channels, word-of-mouth marketing thrives on authentic spontaneous customer reactions. The emergence of digital technology has increased the reach and speed of word-of-mouth, enabling rapid dissemination of information in global networks.

The first link in the chain of communication of innovation involves the innovators. The source of recommendation of an adopted technology comes from various sources. One such scenario occurs when an innovator voluntarily informs potential early adopters about their investment in a technological innovation, driven by their satisfaction and excitement about the innovation. Secondly, the innovator may feel a sense of pride and uniqueness as an early tester of the innovation, which may motivate them to share their investment with others. This can help to reassert their status among peers. Thirdly, the innovators may have the well-being of others in mind. As a result of their positive experience, they may encourage others to replicate his decision to adopt the innovation. Unfortunately, the innovators may also have a negative experience because the innovation has not responded to their needs. Then, they may feel the need to talk about it as a manifestation of dissatisfaction and self-regulation of frustration levels, or to warn others not to replicate their erroneous decisions (Engel et al., 1969).

In the successive stages of innovation diffusion, the process of information dissemination accelerates as more people become involved. The motives for sharing information may also change over time. For example, when representatives of the late majority recommend a product, their motivation for sharing information on their investment made may be to demonstrate membership in the group or to express gratitude to others for their support in adopting the innovation. Interestingly, word-of-mouth may intentionally be the main channel for conveying information. It has been observed that some companies adopt a sales strategy that relies on an informal distribution network built on trust between the supplier, technology service provider, and the customer. Through this planned cooperation, which utilises the relationship between the user and the manufacturer, small companies had the opportunity to learn quickly by using consumer experience and efficient feedback (Van Est, 1999).

Research also suggests that establishing trust between suppliers and customers through a coordinated advisory system strengthens not only practical knowledge, but also user satisfaction with the technology. This, in turn, can lead to increased engagement in promoting the benefits of the solar system to potential downstream users. Recognizing the power of word-of-mouth communication, experienced providers incorporate it into their strategies to enhance brand awareness, increase credibility, and boost sales. They cultivate positive customer relationships, encourage user-generated content creation, and promote through referral programs and partnerships with distributors. This phenomenon is particularly visible in smaller towns, where satisfaction with a specific system such as solar water heaters, for example, can quickly generate the spread of positive word-of-mouth feedback about the system. In smaller towns, supplier-customer relationships are often established on both a social and personal level.

This effect is also seen in the case of ineffective relationship building between supplier and customer. If the supplier fails to provide detailed technical explanations relevant to the operation of the solar heating system and, as a consequence, the system may not work efficiently. This can result in users focusing more on its shortcomings and spreading negative opinions about it (Elmustapha et al., 2018). Negative word-of-mouth can quickly spread, damaging the brand's reputation and undermining trust. Furthermore, in an era of information overload, attracting and retaining consumer attention amidst the deluge of content poses a significant challenge. However, these challenges also present an opportunity for producers and distributors of energy systems to authentically engage in consumer relationships, proactively address concerns, and transform critics into advocates.

The decision to adopt an innovative energy technology involves the acquisition of a certain experience. This experience, depending on whether it is positive or negative, contributes to the way in which the user communicates information to their environment. This, in turn, influences subsequent potential users and their decisions on whether to adopt the innovation in question. The likelihood and effectiveness of the word-of-mouth effect is therefore influenced by satisfaction with the product, trust in the recommender, the relevance of the information and the social ties between individuals. In addition, the emotional tone conveyed, the perceived expertise of the person recommending the innovation, all of which can enhance the impact of word-of-mouth messages, play a role. An attempt to better understand this complex relationship between the many actors in the energy market was undertaken by Michael Mutingi (2013), who developed a model for the adoption of renewable energy technology in a systems dynamics approach. The aim of this study was to answer the question of how the relationships between the multiple actors involved in the diffusion process of an energy innovation impact its diffusion and adoption.

The model developed by Mutingi (Figure 9) depicts a causal loop diagram of the interaction between word-of-mouth and the adoption of energy innovations (Mutingi, 2013, p. 181). As proposed by the author, the interactions occur in three main loops. The first loop in the model is a balancing loop, which is employed to represent the relationship between the number of potential adopters of renewable energy-based technology and the adoption rate. This loop is denoted by B1. The higher the number of potential adopters expressing interest in adopting the technology, the higher the value of the adoption rate. As the adoption rate grows, resulting in fewer potential users who have not yet adopted the technology. Loop B1 explains the market dynamics of a given technological innovation. Promotional activities can further boost the adoption rate and enhance the technology's popularity.

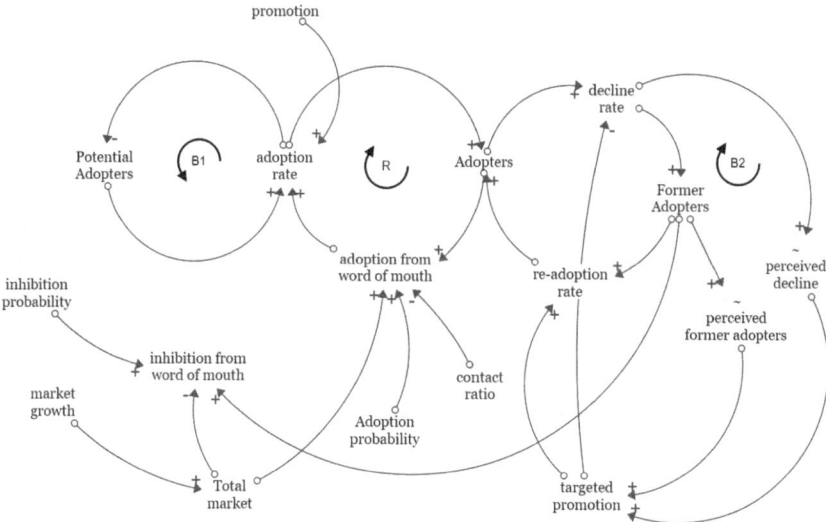

Figure 9. Causal loop analysis for renewable energy adoption, (Mutingi, 2013, p. 181)

Another loop in the model is the amplification loop labelled R. It represents the relationship between user population and adoption rate. As the adoption rate grows, the number of people who have adopted the innovation increases. In turn, the larger the population of active users, the greater the spread of information about the technology in the society. Following the increased flow of information, more people learn about the innovation and decide to adopt it. Furthermore, the effect of the pilfer post depends on the size of the market; the larger the market (total market index), the more significant the spread of information. If the public is receptive to innovation, indicated by a higher probability of adopting the innovation in the near future (Adoption probability index), consumers are more likely to positively and willingly discuss the new technology with others. Conversely, the contact rate diminishes the impact of the word-of-mouth effect. This is because when a larger proportion of the population is already familiar with the technology through direct contact, there is less necessity to discuss it.

The last loop in the model, labelled B2, is a balancing loop that illustrates the relationship between the population of former adopters and the rate of decline in innovation adoption. As the user population increases (the Adopters rate), the decline rate, reflecting dissatisfaction with the technology, also increases. The higher the decline rate, the higher the perceived decline and the larger the population of former users who have abandoned the innovation. Consequently, as perceived decline and former adopters increase, manufacturers and distributors intensify efforts to reach these former users through targeted promotion (targeted promotion rate) and encourage them to reconsider adoption (readoption rate). Simultaneously, increased targeted promotion activities contribute to lowering the decline rate (Mutingi, 2013).

Mutingi employed the developed model to test four different scenarios: (1) Ideal case – no decline in technology use, no re-adoption and no corrective policy control; (2) Adoption with decline – possibility of decrease in technology use, but without re-adoption, without corrective policy control and without influence of former adopters; (3) Adoption with decline and inhibition – possibility of decrease in technology use, negative influence of former adopters but without re-adoption and without corrective policy control; (4) Adoption with fuzzy policy control (Figure 10).

Table 2 shows the detailed simulation results obtained by the author considering the different scenarios.

Table 2. A summary of comparative simulation results, (Mutingi, 2013, p. 192)

Performance criteria	Ideal case	With decline	With decline and inhibition	With fuzzy policy control
Peak adoption rate	8	8	7	8
Peak adoption time	11.26	11.26	10.41	11.26
Maximum adoption	100	85	80	97.5
Maximum adoption time	24	24	16	25

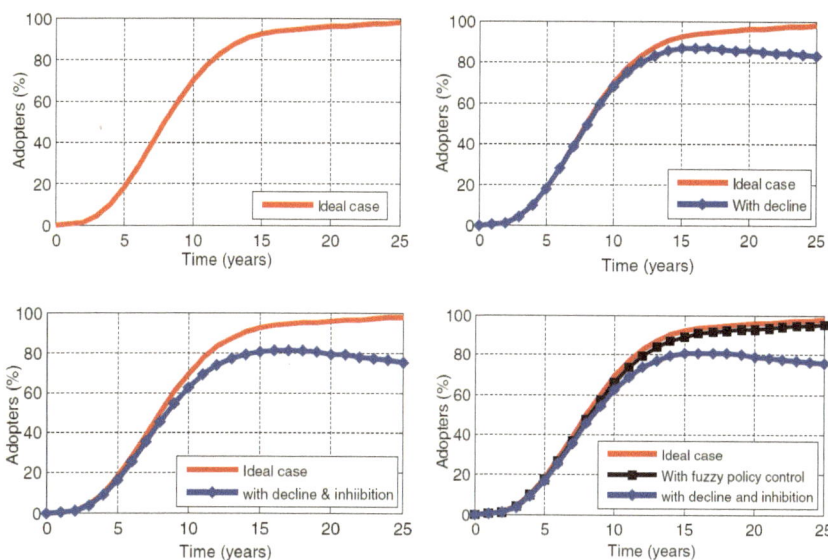

Figure 10. Four scenarios of Renewable Energy Technologies adoption behaviour, (Mutingi, 2013, pp. 188–191)

Word-of-mouth influences the behaviours of potential users of innovative technology at every stage of the process related to its purchase, installation, and use. During the awareness stage, positive recommendations can spark interest and encourage consideration of a particular offer. In the evaluation phase, word-of-mouth acts as a credible source of information, aiding prospective users in making informed decisions about adopting a specific energy solution. After purchase and installation, it reinforces satisfaction and loyalty, potentially leading to advocacy and further diffusion of innovation. Word-of-mouth continues to be a powerful force in the contemporary landscape of social learning utilised in the energy market, driven by the authenticity and credibility of personal recommendations. Its ability to influence perceptions, decisions, and enhance brand loyalty underscores its potential and enduring significance. With technology advancing and consumer behaviours evolving, leveraging the power of word-of-mouth will continue to be a cornerstone of effective marketing strategies. This approach supports authentic connections between producers, suppliers, and users in an increasingly interconnected world.

Chapter 4.
Mechanisms of Peer Effects Formation: Members of Social System

When examining the mechanisms that underlie peer effects, it is crucial to consider the derivative of the word-of-mouth effect. Direct conversations among members of social networks serve as primary sources of information and knowledge regarding renewable energy sources and their utilisation. Conversely within these networks, individuals who are well-known and respected can significantly amplify efforts toward social learning and accelerate the diffusion of innovation. The role of opinion leaders in the innovation diffusion process constitutes a significant area of research in both social sciences and economics.

Opinion Leaders

Opinion leaders are frequently described as individuals or groups who hold authority and possess the capability to influence the opinions, attitudes, and behaviours of others within their environment. Definitions of opinion leaders vary depending on the research perspective and the social or cultural context. One approach to defining opinion leaders focuses on their knowledge and expertise. For instance, Elihu Katz, Paul F. Lazarsfeld, and Elmo Roper (2017) characterise opinion leaders as individuals who possess knowledge or experience in a specific field, are recognized as experts within it, and demonstrate capabilities that exceed those they influence. Such perceived opinion leaders are sought after by others for advice and knowledge sharing.

Due to their experience, knowledge, or authenticity, opinion leaders are trusted by others. Their support for a particular innovation can contribute to building trust and reducing uncertainty associated with its adoption, hence research is conducted on the effectiveness of marketing strategies based on opinion leaders, as well as on methods of identifying and engaging these leaders in promotional activities. When opinion leaders endorse specific solutions related to energy transformation, others are more likely to trust those recommendations and take actions towards sustainable energy. The endorsement of a specific technical solution by opinion leaders is particularly impactful when they actively use it and express satisfaction with its utility, quality, or specific benefits. They contribute to building a positive image of these innovations within the community they identify with. This mechanism is exploited by producers who, especially in the initial stages of introducing a solution to the market, offer it for free to celebrities, bloggers, or other influential individuals in exchange for active product endorsement.

While Cheng Ming Yu defines opinion leaders as "those individuals whose beliefs, practices and behaviours are noticed and imitated by others" (Ming Yu, 2002, p. 80). According to Parau et al. (2017, p. 157), "opinion leaders are individuals who exert a significant amount of influence within their network and who can affect the opinions of connected individuals". Rogers focuses on the temporal aspect, crucial in the innovation diffusion process, proposing that "opinion leadership is the degree to which an individual is able to influence informally other individuals' attitudes or overt behavior in a desired way with relative frequency" (Rogers, 1983, p. 331). Additionally, apart from opinion leaders, Rogers distinguishes change agents. According to his definition, "a change agent is an individual who influences clients' innovation decisions in a direction deemed desirable by a change agency. In most cases a change agent seeks to secure the adoption of new ideas, but he or she may also attempt to slow the diffusion process and prevent the adoption of certain Innovations" (Rogers, 1983, p. 331).

Opinion leaders are often among the first to accept new products, technologies, or ideas. Rogers' theory of the diffusion of innovation (1983) emphasises the importance of leaders as early adopters of innovation, hence research is conducted on the role of leaders in social processes such as attitude change, consumer behaviour, or decision-making processes. Opinion leaders are often perceived as catalysts for social change, influencing the opinions and actions of others in their environment (Valentine, 2010). Their enthusiasm and active adoption of innovations make them more visible and draw the attention of others in the community. They often have a keen eye for new developments in their field of activity, enabling them to quickly identify emerging ideas, products, or technologies emerging on the market. Thanks to their ability to track trends and analyse changes, they can quickly recognize innovations with potential for further development and diffusion. Opinion leaders not only identify new innovations but also actively engage in testing, using, or experimenting with them. Due to their curiosity and openness to new experiences, they are often among the first ones to try out new products or technologies, even if they are still relatively unknown or risky.

Opinion leaders possess the ability to raise social awareness through their actions and communication. Whether through publications, public appearances, participation in social media, or other forms of communication, they can bring attention to new innovations and catalyse discussions about them within the community. Opinion leaders often serve as guides and advisors in the decision-making process regarding the adoption of innovations. Their recommendations and positive opinions can significantly influence others to take action. Openness to change and willingness to experiment can inspire others in the community, encouraging them to explore and adopt new solutions. By utilising authentic voices of satisfied or dissatisfied customers, distributors can support communities of ambassadors and influencers (Ferreira et al., 2022), who promote their products or services. These ambassadors often focus on collecting opinions and feedback from the social network in which they operate. This information can provide valuable insights to distributors on how to improve the innovation or how to tailor its promotion to the needs and preferences of the audience.

Research on the communication channels used by opinion leaders highlights the creativity exhibited by these leaders, particularly with new technologies. Opinion leaders leverage various platforms to maximise their reach and influence, effectively disseminating diverse information about energy innovations within their communities. In the literature, two main categories of opinion leaders can be distinguished: social leaders -trusted and authoritative within their society or social group through the use of traditional communication methods, and media leaders, who achieve their position through activity in mass media. Differences between these two types of leaders can influence their role and effectiveness in the innovation diffusion process.

Taking into account the above division, research encompasses the analysis of various communication channels, such as social media, blogs, public appearances, interviews for traditional media, or participation in social events. Leaders are often also recognized as fashion icons or trendsetters in their community. Their decisions regarding the adoption of new products or technologies can influence others and lead to the emergence of trends or consumption patterns. Researchers analyse how social media and other digital platforms change the dynamics of innovation diffusion and how opinion leaders use these new tools for communication and societal influence (Wang, 2024).

Peer Pressure Effects

The peer pressure phenomenon, also known as social pressure, peer influence, or social influence, refers to a situation in which a peer group exerts influence on an individual to conform to the norms, values, behaviours, or expectations prevailing within that group, adopt a specific stance, or take particular actions (Meshram, 2016). Peer pressure is a complex phenomenon influenced by several key components, three of them playing a crucial role in its effectiveness: (1) the individuals exerting pressure, (2) the individuals subjected to pressure, and (3) the pressure exertion strategy. Peer pressure can be applied in various contexts such as advertising, marketing, politics, education, or daily social interactions. It is essential to understand the goals of the individuals exerting pressure aim to achieve and the motivations, needs, and values of the pressure recipients (Aronson et al., 2005).

The effectiveness of peer influence undoubtedly depends on who exerts the pressure. As described in earlier sections of this monograph, a peer group can consist of different individuals. Due to personal traits, social position, or relationships, the communication between the individuals exerting pressure and the recipients varies. Research suggests that concerning technology adoption, the person conveying information or educating is as important as the information itself or the knowledge resource being conveyed. Interestingly, peers such as family, friends, or other trusted network members are more likely to disseminate information, while government officials or local authorities typically exert social influence (He et al., 2022).

Individuals with high social competencies, capable of gaining the public's trust, are more likely to influence others than speakers who lack social trust (Lin, 2019). Moreover, individuals perceived as more physically attractive, due to personality attributes or communication style, such as gesture mannerisms or vocal intonation, are more successful in gaining their peers' trust (Niebuhr et al., 2023). The potential of such individuals is utilised in the information diffusion process, where they play the roles of opinion leaders or change agents discussed earlier.

The effectiveness of peer influence also depends on the characteristics and preferences of its recipients. Both age, gender, education level, socio-economic status, environmental values, and knowledge as well as less obvious factors that may influence the relationship between the influencer and the recipient of peer pressure are important (Dowd et al., 2012). One of these factors are cultural differences and their associated peer influence through diverse argumentative styles. For instance, for Americans, who prioritise personal preferences, a better way to encourage pro-environmental actions is by using arguments related to personal well-being or individual gain resulting from these actions. Conversely, in Asian countries where collectivist culture prevails, pressure utilising arguments related to the common good and following a common pattern is more effective, meaning it is worth doing something for the environment because others are doing it (Huang et al., 2022).

Considering peer pressure, the following effects can be distinguished. Firstly, the goal of exerted pressure may be to change the recipient's attitudes toward a particular issue, product, or idea, including energy transformation. It may involve convincing the recipient to adopt a positive attitude or acceptance of a given idea, as well as changing negative or reluctant attitudes hindering implementation of pro-environmental practices. An example could be exerting peer pressure to change attitudes toward energy consumption. Studies conducted by Anna Laura Pisello and her team (2016) showed that despite identical characteristics of office rooms (size, shape, natural lighting), peers of comparable characteristics such as age, education, similar routines, and work schedules, exhibited divergent attitudes toward controlling energy consumption. In some office rooms employees maintained a higher temperature, under the same weather conditions, employees in another room regularly maintained a lower temperature. The authors suggest that this is related to diverse perception of thermal comfort or the need to turn lights on and off (Pisello et al., 2016).

Another goal of peer pressure may be to change the recipient's beliefs about a specific issue or product. Hajarini et al. (2022) distinguish three types of beliefs crucial in shaping energy transformation policies. The first group includes behavioural beliefs such as the perception of energy transformation by system users, knowledge and understanding of the energy transformation process by energy users, living conditions and status in the place undergoing transformation, and a sense of justice and rationality of the planned transformation. The second group consists of normative beliefs, which include the need for involvement of various stakeholders and managers of energy transformation. This group also emphasises social cohesion, represented by the regularity and quality of interactions and relationships within the local community, and the mutual influence the community has on the behaviours of energy users. The third set of beliefs comprises control beliefs, which encompass several key aspects: understanding the necessity to establish a framework of representatives serving as social intermediaries on behalf of the community, defining property rights by the state, engaging the local community in the collaborative process of energy transformation to empower them, ensuring their agency, believing in the financial capacity of users to implement the transformation, providing educational readiness for users regarding energy transformation, and guaranteeing users the option to withdraw from the process after it has commenced (Hajarini et al., 2022).

Another goal of peer pressure may be an attempt to persuade the recipient to take specific actions or reactions. This may include purchasing a product, supporting a particular idea, participating in a social campaign, or changing lifestyle or behaviour. For example, Constantine Spandagos and his team explored the role of peer pressure-based educational interventions in shaping energy-saving behaviours among residents of Hong Kong. Based on quantitative survey research, the authors verified two types of interventions, offline interventions, based on direct pressure from the local community, and online interventions, through social media. The results indicate that peer pressure, regardless of how it is exerted, is a significant source of behaviour modification related to heating and cooling practices. Although both types of interventions were highly effective, interactions through social media were stronger (Spandagos et al., 2021).

In business and politics, a crucial objective of applied pressure is often to cultivate a positive brand image for a company, organisation, or public figure. Through appropriate messages and actions, peer pressure can influence society's perception of a particular individual or institution. Peer pressure can also serve to reinforce existing beliefs or attitudes of the recipient. In such cases, the goal is to maintain customer loyalty, strengthen emotional bonds with the brand or idea, or enhance social engagement. An example of such actions is a company's investment in innovative technological solutions and transition to renewable energy sources for dual purposes. This initiative serves dual purposes: firstly, to enhance the company's reputation as an environmentally friendly entity and align with current technological trends; and secondly, to retain existing customers and attract new ones who perceive such actions as socially and environmentally significant (Issa & Hanaysha, 2023).

In some cases, actions based on social influence mechanisms primarily aim to educate recipients about specific issues, social problems, or the benefits associated with a particular product or service. In other instances, this education may occur unconsciously, happening simultaneously. These goals are achieved through selected peer influence strategies. It can be presented in the form of technical facts supported by values or numerical simulations. This is a type of logical persuasion, influencing based on the presentation of rational arguments and evidence to convince the recipient of a particular stance. In this case, the effectiveness of the influence depends on presenting facts, statistics, or scientific research in a way that makes the ensuing arguments understandable, convincing, and based on credible sources (Gosnell et al., 2019).

Another strategy is emotional persuasion. It involves evoking strong emotions in recipients, such as fear, joy, empathy, or anger, to encourage them to take action or change their attitudes. Emotions can be utilised to authenticate the message, reinforce persuasion, or evoke empathy. For instance, the need for eliminating coal-fired boilers can be presented by referring to statistics of illnesses and images of damaged lungs due to breathing polluted air, appealing to the emotions of information recipients. An example of such information dissemination is employed by the Polish Smog Alert organisation as part of the anti-smog campaign "See What You Breathe" (Polski Alarm Smogowy, 2020). Emotional campaigns also frequently utilise celebrities, expert endorsements, industry figures, or influential individuals who reinforce the message with their authority. This mechanism has been detailed in the previous section (Opinion Leaders).

The next strategy for peer influence utilises a technique based on presenting practical examples illustrating the desired effect or indicating that other people also support specific beliefs or behaviours. For instance, during informal conversations or targeted discussions, opinions or referrals from other consumers may be cited to convince recipients to purchase a product. To reach a wider audience of potential adopters, both solar energy vendors and decision-makers employ unconventional communication methods based on presenting recommendations from family, friends, or colleagues who have had positive experiences with installing solar heating systems. These pragmatic marketing tools utilising the peer influence mechanism instil greater trust in renewable energy installations than basic brochures or informational campaigns (Pudaruth et al., 2017).

Another employed strategy used to increase the diffusion of innovations involves addressing potential counterarguments in the communication. Within this strategy, individuals exerting peer influence consider foreseeable or previously diagnosed counterarguments or objections from recipients and attempt to address them in their communication, pre-empting and mitigating potential negative attitudes of the recipient. Demonstrating that various viewpoints have been considered and responding to potential objections may increase the credibility and effectiveness of persuasive efforts. Research in the field of electricity generation technology suggests that presenting arguments consistent with the recipient's initial attitude is perceived as more convincing than those inconsistent with their initial attitude. Presenting a specific technological solution as addressing consumer concerns enhances the likelihood of innovation adoption, especially since authors argue that the tendency for argument consistency with presented attitudes seems to exert the greatest influence on respondent evaluations (Shamon et al., 2019).

The subsequent strategy utilises verbal visualisation techniques by creating narratives and stories facilitating the acceptance of new products or technologies. Storytelling or presenting unconventional narratives can be an effective persuasive strategy because it helps recipients better understand and identify with the conveyed information. Stories can be used to evoke emotions, illustrate abstract concepts, or demonstrate the benefits associated with specific actions. Similar functions are fulfilled by rhetorical techniques such as metaphors, analogies, rhetorical questions, or the use of imagery, enabling increased effectiveness of the message. Storytelling based on individual experiences related to the applied heating system serves as a means of understanding, communicating the adoption process of a given technological solution, and simultaneously influencing peers (Goodchild et al., 2017).

Another strategy used within peer pressure is leveraging arguments based on presenting time constraints or limitations in solution availability. Persuasion can be more effective if the presentation emphasises the need for making a decision within a specified time. Presenting an offer as available to a certain number of people can increase the sense of urgency or rarity of the offer, which in turn may encourage recipients to act more quickly or make decisions to adopt innovations. An example could be arguing for the attractiveness of investing in a photovoltaic installation at the time of changing the method of individual energy production settlement from net metering to net billing (Trela & Dubel, 2021).

Another strategy is based on the principle of consistency and striving to maintain consistency in actions and decisions. Meaningful engagement of peers in the persuasion process, such as through small initial steps, can increase the likelihood of subsequent engagement in larger actions. An advanced installation does not necessarily have to be implemented all at once; related adoption processes in households can be based on acquiring one renewable energy source and then adding another renovation or energy technology, without a stable end point (Juntunen, 2014).

Chapter 5.
Application of Peer Effects: Educational Interventions

Another way to foster the environment for peer effects to emerge is through planned educational initiatives. Resistance to energy transition or slow pace in its implementation may stem from a lack of education and awareness regarding the necessity to change energy production and consumption patterns. People may be unaware of the consequences of climate change or not realise the potential offered by sustainable energy sources. To address this issue, various educational initiatives are undertaken, both grassroots and planned within local, national or international policies. Organised training, raising social awareness, increasing public participation, and providing public access to information about technological innovations are important communication channels on the path to environmental protection and achieving goals related to global decarbonization. Planned educational activities serve not only to reinforce the positive effect of word-of-mouth but also to leverage peer effects enhancing the effectiveness of educational interventions, increasing their attractiveness, and serving as promotional or informational activities.

Educational activities supporting the diffusion of innovation operate under the assumption that lack of knowledge about available tools to protect our planet lies at the heart of the reluctance to change environmentally unfriendly behaviours. Strategies are developed to enhance the knowledge of potential adopters, aiming to effectively reach diverse audiences, particularly those that are harder to engage.

Reinforcing the Word-of-Mouth Effect

Network structures, such as groups, communities, or organisations, play a crucial role in disseminating information and supporting the innovation diffusion process, thus one of the aims of educational activities is to strengthen the functioning of social networks. Creating and supporting network structures, such as communities, organisations, or groups, can facilitate the exchange of information and experiences. Encouraging open dialogue and collaboration between different social groups can promote mutual learning and understanding of diverse perspectives on energy issues.

According to Palm's research (2017), interpersonal networks are effective carriers of innovation knowledge, although the knowledge transmitted in this way is usually not advanced and mainly instructional. Nevertheless, in the conducted study, adopters considered information conveyed through informal contacts valuable as it reduced concerns regarding technological innovation and made them adopt photovoltaic installations. It appears that disseminating information about the reliable and user-friendly operation of photovoltaic systems was crucial. Users providing educational support were perceived as credible sources of information. For prospective adopters, these were individuals they could identify with – people in similar situations with similar needs and familiar acquaintances. The received information about the technology meeting users' expectations regarding system efficiency was perceived as reliable. Through continuous social learning and targeted educational interventions, adopters' knowledge was reinforced, enhancing awareness of environmental issues and understanding of technology operation. The results indicate that active peer effects facilitated through direct interpersonal contact and word-of-mouth mechanisms were stronger within existing social circles of friends, relatives, and colleagues compared to networks of unfamiliar neighbours. Nearly all participants who interacted with an active user of photovoltaic installations before making the adoption decision already knew that person (Palm, 2017).

Building on the findings of Lin and Jia's research (2023), it is recommended that efforts to enhance the word-of-mouth effect through educational interventions should also be integrated in the policy development. The authors emphasise that social networks are an important medium for enhancing peer effects to maximise the benefits of educational campaigns, both for innovation diffusion and other behaviours towards green transformation. The authors recommend undertaking targeted, planned actions addressed to groups that are particularly susceptible to be influenced by peers. The groups in shaping pro-environmental consumer habits include students, organisations based on more personal relationships, groups of environmentally concerned individuals, and individuals with low levels of depression. Universities play a pivotal role in shaping environmentally friendly policies by enhancing training, workshops, and courses that cultivate knowledge and skills related to environmental literacy. They possess the space and potential to promote pro-environmental values and effectively harness peer effects (Lin & Jia, 2023).

Word-of-Mouth to Increase the Effectiveness of Educational Interventions

In the context of formal education, planning additional educational activities or spontaneous educational interventions usually does not present significant challenges. Peer effects are leveraged in two primary ways in this setting. Firstly, educational activities aim to directly support environmental literacy, shape environmentally friendly attitudes among students or beneficiaries. Secondly, peer effects are used to indirectly educate the families of students, following an intergenerational education approach. One of the frequently cited examples of such education is waste segregation. Parents, grandparents, and other members of the family community who need of educational support regarding pro-environmental behaviours receive it from children. In this case, learning often takes the form of experiential learning; children, returning home from school, begin to apply practices learned in school and encourage family members to do the same (Assuah & Johansen, 2023).

Utilising social learning in informal education can be more challenging to implement, but efforts are made to harness word-of-mouth effects to support this form of education. Peer effects play a crucial role in the educational process, influencing not only the expansion of the knowledge base of potential users of technological innovations but also their behaviours or attitudes. Fostering positive interactions among peers can lead to increased motivation to adopt innovations, as well as enhanced skills for effectively using installations, ultimately resulting in satisfaction with the decision made. Encouraging social interactions, fostering a friendly environment among potential adopters themselves, as well as between adopters and local or regional stakeholders, can be achieved by supporting the formation of partnerships for energy transition. Such partnerships have the potential to widely disseminate knowledge about technology adaptation processes, technical aspects related to their use, and exchange of experiences among various actors in the energy market. These include environmental organisations, charities, energy system distributors, energy companies, and local authorities. Through such supported innovation diffusion, potential adopters can benefit from education based on social learning, thereby potentially reducing inequalities in access to green energy (Zhang et al., 2023).

Providing access to education and training on new ideas and practices can increase awareness and acceptance of innovations in society. Encouraging cooperation and mutual assistance within peer groups can foster increased trust and bonds between peers, thereby enhancing the effective utilisation of the technical capabilities of adopted devices. One example of utilising word-of-mouth effects for educational support is peer assistance within user communities, which is particularly crucial in the early stages of technological innovation development. Learning alongside innovation diffusion is crucial in the green transformation process. Users who have gone through the technology adoption process with the support of social networks declare that learning is an important benefit for them because they had to acquire new knowledge to undergo transformation. The knowledge and skills acquired are the result of peer-to-peer learning processes, where earlier adopters pass on specialised knowledge to subsequent users of innovation (Lähteenoja et al., 2022).

Creating peer effects in educational activities can lead to socially challenging situations. Negative interactions among peers can result in conflicts and may promote inappropriate behavioural patterns. In a model developed by Gyuhwan Kim and Taehwa Lee (2022), they present a phenomenon where radical actions taken by initiators of energy transformation encountered resistance from the local community. Despite educational efforts aimed at addressing social needs related to green transformation, aimed at replacing the conventional energy system, they were not well received. Although the current energy system posed various problems, radical actions by initiators of energy transformation faced social resistance because residents perceived them negatively and mutually supported this attitude. Various strategies employed by initiators to overcome this resistance did not yield the desired effect. Consequently, radical actions failed to induce attitude or behavioural modifications among residents, resulting in a lack of structural changes (see Figure 11). Interactions among "radical actions," "social resistance," "strategies," and "citizen participation" were negative – denoted with a minus sign. (Kim & Lee, 2022).

Differences in social and economic status among peers can lead to inequalities in their relationships and hinder the rapid innovation diffusion. Individuals from lower social and economic statuses may have limited access to education, technology, and information, which can delay the adoption of innovation and pro-environmental practices. Research conducted by Luise Vibrans and colleagues (2023) demonstrates that communities from lower social classes exhibit less positive attitudes towards eco-innovations compared to those from higher social classes. These individuals also exhibit less interest in energy innovations. Disparities in social and economic status influence interpersonal relationships and mutual trust within communities. Individuals of higher status are more inclined to accept and promote innovations. This is partly because differences in social and economic status affect the distribution of resources and benefits arising from innovations, making them more accessible to higher-status individuals (Vibrans et al., 2023). Utilising word-of-mouth effects to reinforce educational interventions can contribute to reducing disparities in access to clean energy among individuals of different socio-economic statuses, mitigating inequalities in peer relationships, and facilitating faster diffusion of innovation.

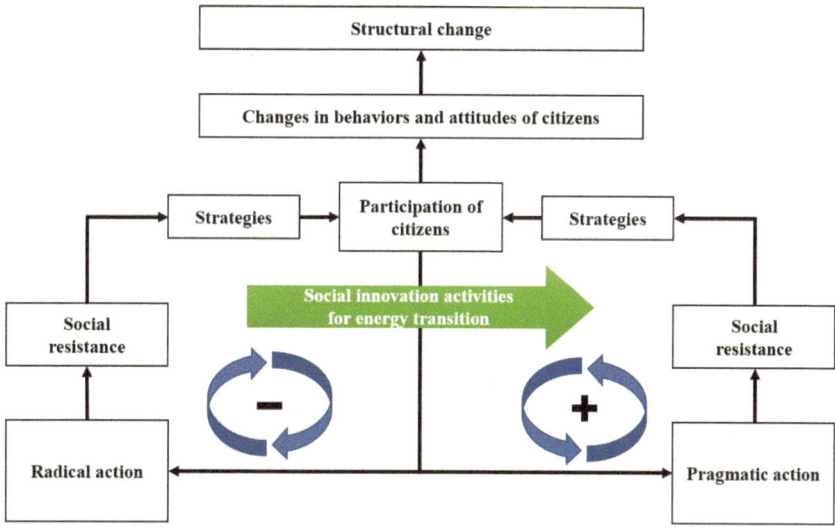

Figure 11. Model of social innovation activities for energy transition (Kim & Lee, 2022, p. 2986)

Promotional Activities

Promotional activities can be implemented in two main ways. Firstly, through direct advertising actions and social campaigns are aimed at promoting the product itself or providing information about available training sessions, workshops, and presentations related to energy transformation. Secondly, the word-of-mouth effect is utilised as an indirect means to persuade consumers to participate in dedicated educational activities. Employing various channels to reach consumers serves increases the reach of educational interventions and strengthens their impact. Alvar Palm and Björn Lantz (2020) conducted research on the effectiveness of informational campaigns in the field of renewable energy. Due to the lag effect, which involves the fact that a photovoltaic installation is not immediately implemented after the informational campaign, investors typically require time to gather the necessary documentation, choose a product, and select a company to install the system. Therefore, the authors analysed adoption trends over a longer period. The analysis covered the period from 2009 to 2016 before the campaign, during which both the experimental and control groups showed a similar trend in the adoption rate, and the year of the campaign itself, 2017 (Figure 12), with the informational campaign taking place in Sweden from April to October 2017. The conducted study demonstrates that running informational campaigns leads in an increase in the adoption rates of photovoltaic installations. During the examined campaign, the number of submitted and approved applications for subsidies for individual installations increased by 29% (Palm & Lantz, 2020).

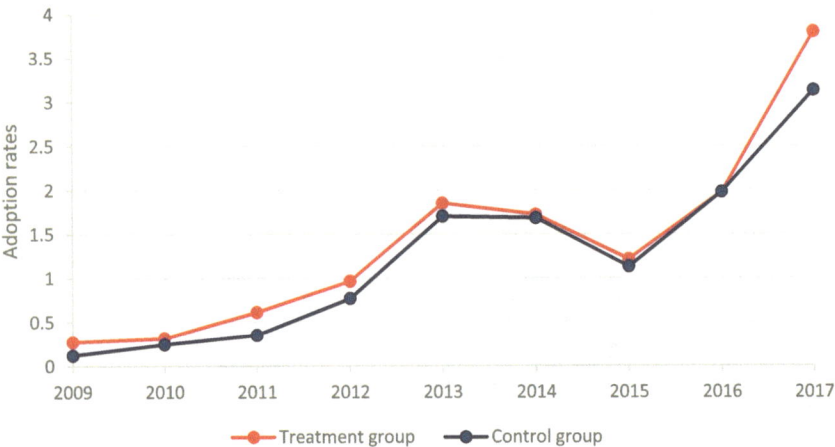

Figure 12. The average rate of photovoltaic installations in residential buildings per 10,000 inhabitants, (Palm & Lantz, 2020, p. 7)

Conversely, word-of-mouth marketing is a strategy that capitalises on the social transmission of information to promote products, services, or ideas. This approach relies on trust and recommendations from third parties, which often makes it more credible and effective than traditional advertising methods. Word-of-mouth marketing can be utilised to promote pro-environmental educational programs, online courses, or social initiatives for local decarbonization. A clear benefit of this approach is the increased reach of innovation diffusion. Word-of-mouth marketing, driven by messages from third parties, can help reach new audience groups interested in technological novelties. It can serve as an interesting and effective promotion strategy in educational activities, but it requires careful planning and execution. It is essential to consider both the benefits and challenges associated with this strategy to achieve positive results in promoting educational programs and social initiatives.

Compared to traditional forms of advertising, word-of-mouth marketing does not require large financial investments, especially when relying on well-functioning and extensive social networks. This is particularly evident in smaller communities. Research shows that individuals within social networks containing a larger number of people knowledgeable about innovations are more inclined to adopt them, irrespective of the number of network members who are active users of energy innovations. Prospective adopters actively seeking information about a particular technology first turn to trusted members of their social network, such as family and friends. This process of acquiring knowledge through social learning is more precise and effective.

Conversely, for more dispersed networks, new media provide an effective alternative. When users share content related to energy transformation on their social networks, this information can quickly spread among their acquaintances, family, and colleagues. In this way, it is possible to achieve a wide reach of communication and promotion of energy transformation. Specifically, word-of-mouth marketing encourages active user participation in the process of disseminating information about available educational offerings by sharing content, commenting, and recommending it to others. This enables the building of social bonds, engages communities in discussions about sustainable energy, and encourages specific educational actions to improve energy efficiency and utilise renewable energy sources. The mechanism of diffusion of innovation by using internet and social media platforms to convey information does not require significant financial investments, yet it can yield significant results in increasing social awareness of energy transformation and interest in participating in related training (Wang & Sun, 2022).

However, employing word-of-mouth marketing also entails certain risks. During multiple transmissions of information among peers, there is a risk of losing control over the message and its interpretation, particularly when the campaign relies on third-party individuals. This can lead to situations where messages are distorted or conveyed in a manner inconsistent with educational objectives or the values of educational support initiators, resulting in misunderstandings or erroneous interpretations. Content sharers may lack awareness or expertise in the energy field, potentially leading to the dissemination of false or misleading information. Disruptions in information flow, resulting in discrepancies between sellers' promises or recommendations and consumer expectations, can create challenging situations, potentially leading to failures in the implementation of innovation contextualization processes.

Unfortunately, word-of-mouth marketing can also be used for manipulating audiences or spreading spam. There is a risk that individuals or organisations with alternative views on environmental issues may exploit word-of-mouth marketing mechanisms to promote false products, services, or ideas related to energy transformation, which could harm the reputation of this field and slow down the diffusion of innovation. Social movements and organisations opposing changes in energy policy and the implementation of sustainable energy sources are particularly active in this area. This occurs for several main reasons, and their arguments and strategies may vary depending on the local, political, and social context.

Some companies operating in the energy sector derive their profits from traditional, conventional energy sources such as coal, oil, or natural gas. Energy transformation requires changes in these business models and a shift towards more sustainable and environmentally friendly energy sources, which may conflict with the interests of these companies. Therefore, resistance from industrial lobbyists is often observed, as they seek to maintain the status quo and delay changes in energy policy. For companies representing sectors such as coal mining or petrochemical industry, transitioning to sustainable energy sources may involve job reductions. Individuals associated with these industries often fear job loss and challenges in retraining for positions in new sectors. This can lead to resistance against energy transformation and increased activity to delay or block the diffusion of renewable energy-based technologies.

Another risk associated with inadequate promotional activities is the creation of an echo chamber effect, where social media users are primarily exposed to content and opinions that reinforce their existing views and beliefs. As a result, individuals who are sceptical about energy transformation may be isolated from messages promoting sustainable energy sources, making it challenging to reach them with information on this topic. These individuals remain within social networks of users with similar views, and social learning in such a situation manifests as an undesired peer effect, creating and reinforcing common beliefs that hinder or even obstruct energy transformation.

The consideration of primary factors like word-of-mouth and advertising in innovation diffusion has found reflection in fields such as economics, marketing, technology management, and system dynamics. The Bass diffusion model is one of the most popular mathematical models describing the processes of spreading new products or innovations in society. Proposed by Frank M. Bass in 1969, it serves as a theoretical foundation for analysing the dynamics of diffusion in the market. It is an analytical tool for forecasting the pace and scope of innovation adoption, as well as identifying factors influencing this process. According to the Bass model, diffusion unfolds over time and encompasses various phases, as described by the author's differential equations. These equations account for both the rate of independent innovation adoption and the rate of adoption under the influence of others.

John D. Sterman (2000) implemented the Bass model within the framework of system dynamics. Sterman's model (Figure 13) integrates feedback loops between word-of-mouth – understood as social exposure and mimicry -and the advertising effect, which acts as an external driver of knowledge, awareness, and adoption.

The Bass model assumes that the diffusion process takes place through two main pathways: innovators influence their environment by encouraging others to adopt the innovation, while later adopters decide to adopt the novelty under the influence of observation of others and social learning and external stimuli -two

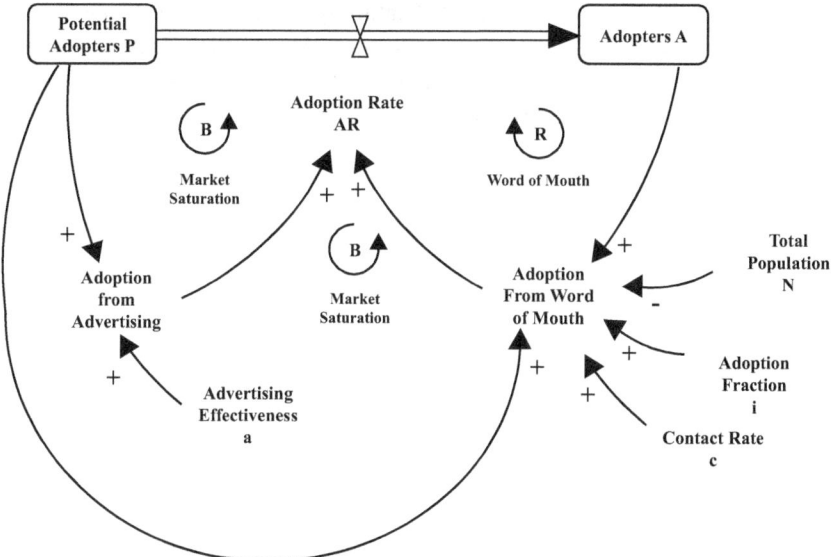

Figure 13. Bass diffusion model (Sterman, 2000, p. 333)

independent sources of innovation. Therefore, the total adoption rate, as described in the AR model, can be represented by relationship (3-1).

$$AR = Adoption\ from\ Advertising + Adoption\ from\ Word\ of\ Mouth \qquad (3\text{-}1)$$

The first stage of the energy transition process involves introducing the innovation to the market. At this initial stage, the innovation is not yet known to the public, so the population of adopters is zero. This means that the only source of adoption will be the first component of equation (3-1) external influences. Adoption at this phase primarily depends on external influences, particularly the effectiveness of advertising (a), which is measured as adoption rates from advertising (3-2):

$$Adoption\ from\ Advertising = aP = \frac{1}{time\ period}P \qquad (3\text{-}2)$$

The impact of advertising is most significant during the initial stage of the innovation diffusion process and diminishes as the pool of potential adopters shrinks. At the same time, the population of adopters increases, leading to the growth of the second component of equation (3-1), adoption through word-of-mouth. This component depends on factors such as the total population (N), the fraction of adoptions (i) and the contact rate (c), according to relation (3-3):

$$Adoption\ from\ Word\ of\ Mouth = \frac{ciPA}{N} = \frac{ciPA}{P+A} \qquad (3\text{-}3)$$

As the user population grows, the pool of naturally diminishes. Consequently, the influence of advertising on adoption rates decreases, while adoption through word-of-mouth increases. This pattern aligns with the logistic model of diffusion, as illustrated in Figure 6 (Sterman, 2000).

Chapter 6.
Modelling Peer Effects

Peer effects as a mechanism of social learning for the diffusion of green energy technologies have attracted the interest not only of researchers dealing with pedagogical or social issues but also of researchers dealing with modelling social relations and interactions also of the researchers using tools of system dynamics play a significant role. This chapter presents examples of modelling and analytical approaches to the phenomenon of peer effects in the context of green transformation. Due to the research context of this monograph, the focus has mainly been on European examples.

Merla Kubli from Switzerland should be mentioned among European researchers modelling the adoption of individual energy systems. The researcher, along with co-authors, has presented successive model proposals considering social learning as a learning effect (see Figure 14) and in the subsequent model as a peer effect (see Figure 15).

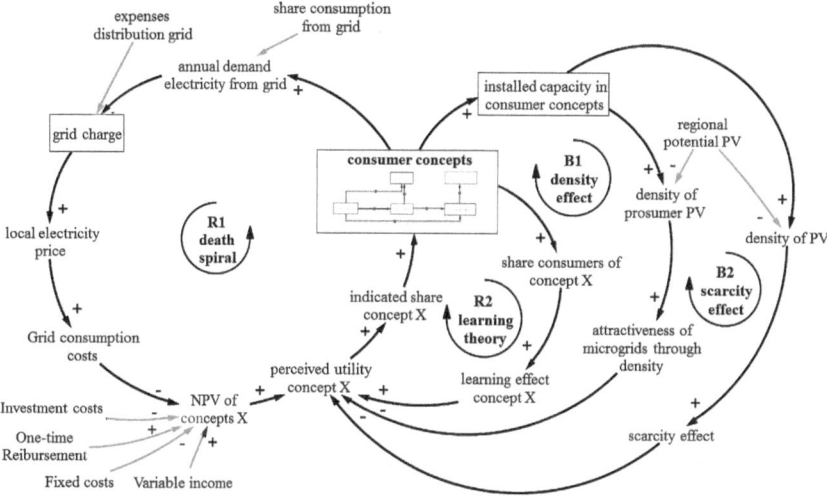

Figure 14. Model with representation of feedback loops (Kubli & Ulli-Beer, 2016, p. 75)

According to M. Kubli's proposition (Kubli & Ulli-Beer, 2016; Kubli, 2018; Zapata et al., 2019), peer-interaction-based learning constitutes a significant component in perceiving the benefits of individual energy generation as a self-consumption concept (Bollinger & Gillingham, 2012). Therefore, in the model of the technological innovation diffusion, it has been incorporated into the peer effect feedback loop (see Figure 15). This feedback loop depicts the investor's decision to acquire a photovoltaic installation as a result of the positive influence of neighbours who own a similar installation. Consumer awareness and knowledge levels regarding renewable energy-based systems increase with a greater number of installed systems in individual households. In this model, therefore, the authors assume that a higher level of awareness resulting from exposure to advertising and the visibility of neighbouring installations, as well as a higher level of information acquired in social learning, leads to a more positive evaluation of the technological solution.

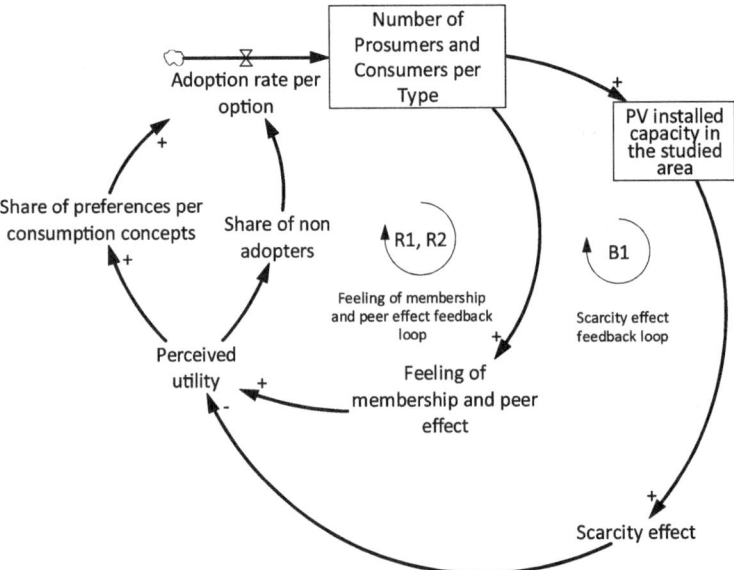

Figure 15. Fragment of M. Kubli's model (Zapata Riveros et al., 2019, p. 3)

In M. Kubli's model, the learning effect (alternatively, the peer effect) is determined based on a constant variable, employing the concept of the peer effect coefficient proposed by B. Bollinger and K. Gillingham (2012) according to the linear relationship (4-1):

$$learning\ effect_X = (share\ consumers\ of\ concept\ X \times learning\ effect\ coefficient) + 1 \qquad (4\text{-}1)$$

The methodology for determining the learning effect coefficient of 0.78, as described earlier in this chapter, used the location of homes with PV installations in the US determined by postal codes (Kubli & Ulli-Beer, 2016).

As Merla Kubli and Sanket Puranik suggest "the network effect design option "peer and community effect" highlights the fact that innovations not only spread through their pure functionality and cost benefits, but also through the way social processes play a role in innovation diffusion. People tend to get inspired by their peers and only for this reason evaluate a product differently. Belonging to a community can be a powerful feeling. An energy community can create a community feeling that awakens the desire to join this group of people. Network effects can cause a lock-in effect, where people enjoy the social benefits of the energy community and are reluctant to change to new/other solutions" (Kubli & Puranik, 2023, p. 8)

Another model considering social learning during the decision-making process regarding the adoption of solutions based on renewable energy sources was presented by Sujeetha Selvakkumaran and Erik O. Ahlgren (2018) and developed for the Swedish market. This model defines the adoption rate as a function of interaction between potential households (single- or multi-family residential buildings) where photovoltaic systems have not been installed and households that have adopted photovoltaics (Figure 16), also treated as communication channels in the model.

The presented model is theoretical, as the authors did not conduct simulations, which would allow to perform a quantitative analysis of the relationships between individual variables. The researchers adopted a bottom-up approach, considering a specific case of the local energy transformation process (case study), and then attempting to generalise it using a hybrid approach. The aim of such an approach was to gain a complete understanding of social interactions in the process of co-creating local energy transformations, rather than their quantitative dependencies within the diffusion of photovoltaic solutions (Selvakkumaran & Ahlgren, 2018).

An interesting observation from the analysis of existing system dynamics models dedicated to the adoption of solutions based on renewable energy sources is that learning is often depicted as a supporting factor in the diffusion of innovation. In the model by Milad Mousavian et al. (2020), they define the learning effect loop as reinforcing (see Figure 17 and Figure 18).

According to the authors' proposal, an increase in installed capacity leads to greater knowledge and experience in using renewable systems. This reduces the cost of capital, thereby accelerating return on investment. Simultaneously, higher returns on investment reinforce the willingness to invest in renewable resources, resulting in more applications for investment funding, an increase in the number of investments, and subsequently, the cumulative installed capacity of devices (Mousavian et al., 2020).

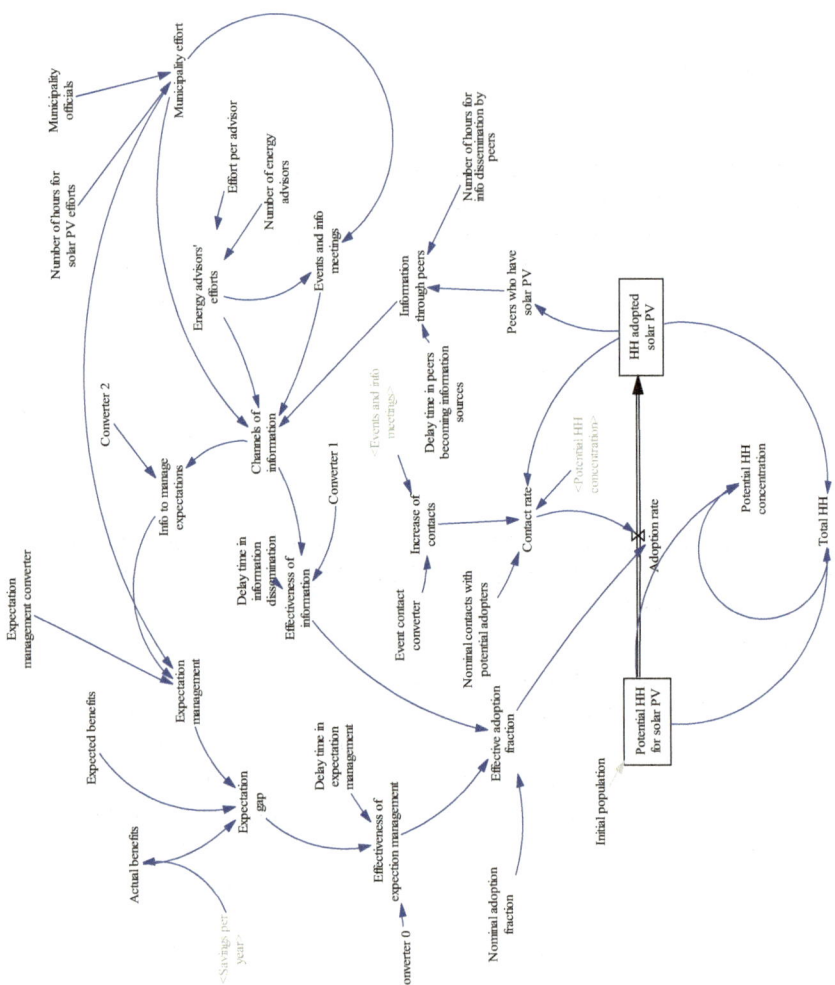

Figure 16. The model proposed by S. Selvakkumaran i E.O. Ahlgren (2018, p. 11)

Modelling Peer Effects

Figure 17. The model proposed by M. Mousavian et al. (2020, p. 1256)

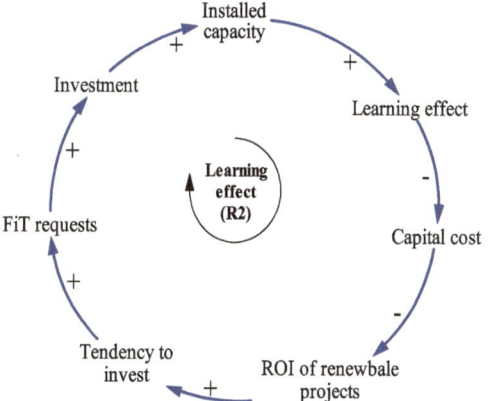

Figure 18. The learning effect loop proposed by M. Mousavian et al. (2020, p. 1255)

Conclusions

Social learning, as demonstrated in this monograph, occurs in diverse ways, evolving with social changes and technological advancements, as communication channels among people also evolve. Introducing technological innovations to the market, especially those aimed at environmental protection and combating or mitigating climate change, requires not only educating society but, above all, planning these actions in a more comprehensive manner.

The need for raising awareness in society applies to both consumers and producers, distributors of technological innovations, and, importantly, local or national authorities who should consistently include well-planned educational activities in local policies. Responsibility for environmental education lies on both sides of these actions. Understanding the processes of social learning, which lead to greater acceptance of environmentally friendly solutions, as well as those that hinder the implementation of positive changes, significantly impacts the quality of our lives. Unfortunately, we still contend with conflicting interests among various groups of stakeholders, including potential conflicts of interest between them. Therefore, education and environmental literacy play a crucial role in ensuring sustainable development and achieving their goals (Sinha et al., 2020).

This monograph emphasises the role of social learning in the process of green transformation because even the best innovation cannot be implemented if it does not gain people's acceptance. Considering social factors, including the multi-level process of learning, is a significant element of planning social change. The content presented in individual chapters will therefore hold significant value for educators planning interventions based on mechanisms of social learning, engineers designing innovative solutions, and individuals undertaking actions for their implementation. The importance of education and pedagogical practice is particularly emphasised here.

Importantly, the monograph is intended to support systems dynamics modelling, a tool used to aid decision-making process related to local decarbonization by visualising the potential effects or consequences of adopted choices. The tool has a supporting role by presenting possible scenarios in a somewhat simplified manner, based on an estimation of the scale of individual effects. It is not a substitute for the deeper reflection necessary in such situations. As a mathematical tool, it aids in estimating economic, social or environmental effects, facilitating key decision-making. However, it does not encompass reflection on the complex goals or competing interests of individual groups in society, such as the economic gains of intermediaries or distributors of pro-environmental technological innovations, or the political gains associated with local authority camps and their public policies, including the economic benefits accruing to these authorities. This tool can thus be seen as a socio-technical instrument. It is important to approach it with caution, while acknowledging its educational value and the potential for widespread implementation in various educational contexts.

References

An, W. (2011). Models and methods to identify peer effects. In J. Scott, & P. J. Carrington (Eds.). *The Sage handbook of social network analysis* (pp. 514–532). SAGE.

Arimura, T. H., Katayama, H., & Sakudo, M. (2016). Do Social Norms Matter to Energy-Saving Behavior? Endogenous Social and Correlated Effects. *Journal of the Association of Environmental and Resource Economists, 3*(3), 525–553. https://doi.org/10.1086/686068.

Aronson, E., Wilson, T. D., & Sommers, S. R. (2005). *Social psychology.* Pearson Education India.

Aschhoff, B., and Grimpe, C. (2014). Contemporaneous Peer Effects, Career Age and the Industry Involvement of Academics in Biotechnology. *Research Policy, 43*(2), 367–381. https://doi.org/10.1016/j.respol.2013.11.002.

Assuah, A., & Johansen, S. (2023)."They taught the children in school and it was them that came home to teach me" – community perspectives on how students influenced recycling attitudes and behaviours in a remote First Nation in Canada. *The Journal of Rural and Community Development, 18*(1), 118–139.

Baddeley, A. D., Allen, R. J., & Hitch, G. J. (2017). Binding in visual working memory. The role of the episodic buffer. In A. Baddeley (Ed.). *Exploring Working Memory* (p. 312–331). Routledge.

Bandura, A. (1977). *Social learning theory.* Englewood Cliffs, NJ: Prentice Hall.

Bandura, A. (1999). A social cognitive theory of personality. In L. Pervin & O. John (Ed.), *Handbook of personality* (2nd ed., pp. 154–196). New York: Guilford Publications.

Bandura, A. (2021). *Psychological Modeling. Conflicting Theories.* Routledge. https://doi.org/10.4324/9781003110156.

Bass, F. M. (1969). A new product growth for model consumer durables. *Management science, 15*(5), 215–227. https://doi.org/10.1287/mnsc.15.5.215.

Belzer, A., & Dashew, B. (Eds.). (2023). *Understanding the Adult Learner: Perspectives and Practices.* Routledge.

Ben Maalla, E. M., & Kunsch, P. L. (2008). Simulation of micro-CHP diffusion by means of System Dynamics. *Energy Policy, 36*(7), 2308–2319. https://doi.org/10.1016/j.enpol.2008.01.026.

Biresselioglu, M. E., Solak, B., & Savas, Z. F. (2024). Unveiling resistance and opposition against low-carbon energy transitions: A comprehensive review. *Energy Research & Social Science, 107,* 103354. https://doi.org/10.1016/j.erss.2023.103354.

Bollinger, B., & Gillingham, K. (2012). Peer Effects in the Diffusion of Solar Photovoltaic Panels. *Marketing Science, 31*(6), 900–912. https://doi.org/10.1287/mksc.1120.0727.

Brent, D. A., Cook, J. H., & Lassiter, A. (2022). The effects of eligibility and voluntary participation on the distribution of benefits in environmental programs: An application to green stormwater infrastructure. *Land Economics, 98*(4), 579–598. https://doi.org/10.3368/le.98.4.102920-0166R.

Chan, T. Y., Li, J., & Pierce, L. (2014a). Learning from Peers: Knowledge Transfer and Sales Force Productivity Growth. *Marketing Science, 33*(4), 463–484. https://doi.org/10.1287/mksc.2013.0831.

Chan, T. Y., Li, J., & Pierce, L. (2014b). Compensation and Peer Effects in Competing Sales Teams. *Management Science, 60*(8), 1965–1984. https://doi.org/10.1287/mnsc.2013.1840.

Cinelli, M., De Francisci Morales, G., Galeazzi, A., Quattrociocchi, W., & Starnini, M. (2021). The echo chamber effect on social media. *Proceedings of the National Academy of Sciences, 118*(9), e2023301118. https://doi.org/10.1073/pnas.2023301118.

Clark, B. R. (2021). *Adult education in transition: A study of institutional insecurity.* University of California Press.

Collins, K., & Ison, R. (2009). Jumping off Arnstein's ladder: social learning as a new policy paradigm for climate change adaptation. *Environmental policy and governance, 19*(6), 358–373. https://doi.org/10.1002/eet.523.

Commission of the European Communities (2000). *A Memorandum on Lifelong Learning.* Commission Staff Working Paper.

Cook, J. J., Cruce, J., O'Shaughnessy, E., Ardani, K., & Margolis, R. (2021). Exploring the link between project delays and cancelation rates in the U.S. rooftop solar industry. *Energy Policy, 156*, 112421. https://doi.org/10.1016/j.enpol.2021.112421.

Cook, J. J., Jena, S., Qasim, M. S., & O'Shaughnessy, E. (2023). *Observations and Lessons Learned From Installing Residential Roofing-Integrated Photovoltaics.* Golden, CO: National Renewable Energy Laboratory. NREL/TP-6A20-85230. https://www.nrel.gov/docs/fy23osti/85230.pdf.

Dowd, A., Ashworth, P., Carr-Cornish, S., & Stenner, K. (2012). Energymark: Empowering individual Australians to reduce their energy consumption. *Energy Policy, 51*, 264–276. https://doi.org/10.1016/j.enpol.2012.07.054.

Elmustapha, H., Hoppe, T., & Bressers, H. (2018). Understanding stakeholders' views and the influence of the socio-cultural dimension on the adoption of solar energy technology in Lebanon. *Sustainability, 10*(2), 364. https://doi.org/10.3390/su10020364.

Engel, J. F., Kegerreis, R. J., & Blackwell, R. D. (1969). Word-of-mouth Communication by the Innovator. *Journal of Marketing, 33*(3), 15–19. https://doi.org/10.1177/002224296903300303.

European Union (2022). Council Recommendation of 16 June 2022 on ensuring a fair transition towards climate neutrality, (2022/C 243/04).

Fenwick, T., & Tennant, M. (2020). Understanding adult learners. In *Dimensions of adult learning* (pp. 55–73). Routledge.

Festinger, L. (1954). A theory of social comparison processes. *Human Relations, 7*, 117–140.

Ferreira, P., Almeida, D., Dionísio, A., Bouri, E., & Quintino, D. (2022). Energy markets – Who are the influencers?. *Energy, 239*, 121962. https://doi.org/10.1016/j.energy.2021.121962.

Frakes, M. D., & Wasserman, M. F. (2018). *Knowledge spillovers and learning in the workplace: evidence from the US patent office* (No. w24159). National Bureau of Economic Research.

Gillingham, K.T., & Bollinger, B. (2021). Social learning and solar photovoltaic adoption. *Management Science, 67*(11), 7091–7112. https://doi.org/10.1287/mnsc.2020.3840.

Goodchild, B., Ambrose, A., & Maye-Banbury, A. (2017). Storytelling as oral history: revealing the changing experience of home heating in England. *Energy research & social science, 31*, 137–144. https://doi.org/10.1016/j.erss.2017.06.009.

Gosnell, H., Gill, N., & Voyer, M. (2019). Transformational adaptation on the farm: Processes of change and persistence in transitions to 'climate-smart'regenerative agriculture. *Global Environmental Change, 59*, 101965. https://doi.org/10.1016/j.gloenvcha.2019.101965.

Gupta, S., & Ogden, D. T. (2009). To buy or not to buy? A social dilemma perspective on green buying. *Journal of consumer marketing, 26*(6), 376–391. https://doi.org/10.1108/07363760910988201.

Hajarini, M. S., Zuiderwijk, A. M. G., Diran, D. D. D., & Chappin, E. J. L. (2022). Energy users' social drivers to transition from natural gas: a Dutch municipality case study. *IOP Conference Series: Earth and Environmental Science, 1085*(1), 012045. https://doi.org/10.1088/1755-1315/1085/1/012045.

Hampton, G., & Eckermann, S. (2013). The promotion of domestic grid-connected photovoltaic electricity production through social learning. *Energy, Sustainability and Society, 3*, 23. https://doi.org/10.1186/2192-0567-3-23.

He, P., Lovo, S., & Veronesi, M. (2022). Social networks and renewable energy technology adoption: Empirical evidence from biogas adoption in China. *Energy Economics, 106*, 105789. https://doi.org/10.1016/j.eneco.2021.105789.

Heiskanen, E., Happonen, J., Matschoss, K., & Mikkonen, I. (2022). Learning from failures- Encouraging lesson-sharing in the Finnish energy transition. *Energy Research & Social Science, 90*, 102676. https://doi.org/10.1016/j.erss.2022.102676.

Heiskanen, E., Nissilä, H., & Tainio, P. (2017). Promoting residential renewable energy via peer-to-peer learning. *Applied Environmental Education & Communication, 16*(2), 105–116. https://doi.org/10.1080/1533015X.2017.1304838.

Huang, T., Leung, A. K. Y., Eom, K., & Tam, K. P. (2022). Important to me and my society: How culture influences the roles of personal values and perceived group values in environmental engagements via collectivistic orientation. *Journal of Environmental Psychology, 80*, 101774. https://doi.org/10.1016/j.jenvp.2022.101774.

International Energy Agency (2023). Russia's War on Ukraine https://www.iea.org/topics/russias-war-on-ukraine.

Issa, A., & Hanaysha, J. R. (2023). Powering profits: how renewable energy boosts financial performance in European non-financial companies. *International Journal of Accounting & Information Management, 31*(4), 600–622. https://doi.org/10.1108/IJAIM-03-2023-0055.

Jarvis, P. (2004). *Adult education and lifelong learning: Theory and practice*. Routledge.

Juntunen, J. (2014). Domestication pathways of small-scale renewable energy technologies. *Sustainability: Science, Practice and Policy, 10*(1), 1206e1230. https://doi.org/10.1080/15487733.2014.11908130.

Kato, T., & Shu, P. (2009). Peer effects, social networks, and intergroup competition in the workplace. *University of Aarhus, Aarhus School of Business, Department of Economics, working paper no. 09*, 12.

Katz, E., Lazarsfeld, P. F., & Roper, E. (2017). *Personal influence: The part played by people in the flow of mass communications*. Routledge. https://doi.org/10.4324/9781315126234.

Kim, G., & Lee, T. (2023). Understanding social innovation activities for energy transition: Evidence from experiences of social innovation agents in South Korea. *Energy & Environment, 34*(8), 2976–2989. https://doi.org/10.1177/0958305X221116180.

Kristjanson, P., Harvey, B., Epp, M. Van & Tornton, P.K. (2014). Social learning and sustainable development. *Nature Climate Change, 4*(1), 5–7. http://dx.doi.org/10.1038/nclimate2080.

Krot, K., & Lewicka, D. (2013). The market orientation as a key dimension of innovation culture-Study of Polish lingerie company. *International Journal of e-Education, e-Business, e-Management and e-Learning, 3*(2), 79. http://doi.org/10.7763/IJEEEE.2013.V3.197.

Kubli, M. (2018). Squaring the sunny circle? On balancing distributive justice of power grid costs and incentives for solar prosumers. *Energy Policy, 114*, 173–188. https://doi.org/10.1016/j.enpol.2017.11.054.

Kubli, M., & Puranik, S. (2023). A typology of business models for energy communities: Current and emerging design options. *Renewable and Sustainable Energy Reviews, 176*, 113165. https://doi.org/10.1016/j.rser.2023.113165.

Kubli, M., & Ulli-Beer, S. (2016). Decentralisation dynamics in energy systems: A generic simulation of network effects. *Energy Research & Social Science, 13*, 71–83. https://doi.org/10.1016/j.erss.2015.12.015.

Kurantowicz, E. (2012). O społecznym uczeniu się we współczesnym dyskursie andragogicznym. Wątpliwości zebrane. *Dyskursy Młodych Andragogów/Adult Education Discourses, 13*, 13–20. https://doi.org/10.34768/dma.vi13.203.

Lähteenoja, S., Hyysalo, S., Lukkarinen, J., Marttila, T., Saarikoski, H., Faehnle, M., & Peltonen, L. (2022). What does it take to study learning in transitions? A case of citizen energy in Finland. *Sustainability: Science, Practice and Policy, 18*(1), 651–664. https://doi.org/10.1080/15487733.2022.2109316.

Larson, N., Sekhri, S., & Sidhu, R. (2016). Adoption of laser levellers and water-saving in agriculture. *Water Resource and Economics, 14*, 44–64. https://doi.org/10.1016/j.wre.2015.11.001.

Li, Y., Qing, C., Guo, S., Deng, X., Song, J., & Xu, D. (2023). When my friends and relatives go solar, should I go solar too? – Evidence from rural Sichuan province, China. *Renewable Energy, 203*, 753–762. https://doi.org/10.1016/j.renene.2022.12.119.

Lin, B., & Jia, H. (2023). The role of peers in promoting energy conservation among Chinese university students. *Humanities and Social Sciences Communications, 10*(1), 1–10. https://doi.org/10.1057/s41599-023-01682-2.

Lin, W. (2019). Three Modes of Rhetorical Persuasion. *Sino-US English Teaching, 16*(3), 106–112. https://doi.org/10.17265/1539-8072/2019.03.003.

Liu, D., Qi, S., & Xu, T. (2023). Visual observation or oral communication? The effect of social learning on solar photovoltaic adoption intention in rural China. *Energy Research & Social Science, 97*, 102950. https://doi.org/10.1016/j.erss.2023.102950.

London, M. (2011). Lifelong learning: introduction. *The Oxford handbook of lifelong learning*, 3–11. Oxford Library of Psychology.

Manski, C. F. (1993). Identification of Endogenous Social Effects: The Reflection Problem. *The Review of Economic Studies, 60*(3), 531. https://doi.org/10.2307/2298123.

Mastroeni, L., Naldi, M., & Vellucci, P. (2023). Wind energy: Influencing the dynamics of the public opinion formation through the retweet network. *Technological Forecasting and Social Change, 194*, 122748. https://doi.org/10.1016/j.techfore.2023.122748.

McLean, H., Ewart, J. (2020). Managing Relationships. In *Political Leadership in Disaster and Crisis Communication and Management: International Perspectives and Practices*, (pp. 111–130). Palgrave Macmillan, Cham. https://doi.org/10.1007/978-3-030-42901-0_6.

Meshram, R. G. (2016). The upside of peer pressure on adolescent behavior. *Indian Journal of Health and Wellbeing, 7*(8), 845.

Menon, N. M., & Sujatha, I. (2021). Influence of Rogers' theory of innovation of diffusion on customer's purchase intention–a case study of solar photovoltaic panels. *IOP Conference Series: Materials Science and Engineering, 1114*, 012059. https://doi.org/10.1088/1757-899X/1114/1/012059.

Ming Yu, C. (2002). Socialising knowledge management: The influence of the opinion leader. *Journal of Knowledge Management Practice, 3*(3), 76–83.

Mogaji, E. (2021). Brand Integration. In *Brand Management* (pp. 123–144). Palgrave Macmillan, Cham. https://doi.org/10.1007/978-3-030-66119-9_6.

Montrey, M., & Shultz, T. R. (2020). The evolution of high-fidelity social learning. *Proceedings of the Royal Society B, 287*(1928), 20200090.

Moreland, R. L. (2010). Are dyads really groups?. *Small Group Research, 41*(2), 251–267. https://doi.org/10.1177/1046496409358618.

Mousavian, H. M., Shakouri, G. H., Mashayekhi, A. N., & Kazemi, A. (2020). Does the short-term boost of renewable energies guarantee their stable long-term growth? Assessment of the dynamics of feed-in tariff policy. *Renewable energy, 159*, 1252–1268. https://doi.org/10.1016/j.renene.2020.06.068.

Murtonen, M., & Lehtinen, E. (2020). Adult learners and theories of learning. In *Development of Adult Thinking* (pp. 97–122). Routledge.

Mutingi, M. (2013). Adoption of Renewable Energy Technologies: A Fuzzy System Dynamics Perspective. In H. Qudrat-Ullah (eds). *Energy Policy Modeling in the 21st Century. Understanding Complex Systems* (pp. 175–196). Springer, New York. https://doi.org/10.1007/978-1-4614-8606-0_10.

Niebuhr, O., D'Errico, F., Esposito, A., Schmid, E., & Brem, A. (2023). Effective and attractive communication signals in social, cultural, and business contexts. *Frontiers in Psychology, 14*, 1205329. https://doi.org/10.3389/fpsyg.2023.1205329.

Noll, D., Dawes, C., & Rai, V. (2014). Solar Community Organizations and active peer effects in the adoption of residential PV. *Energy Policy, 67*, 330–343. https://doi.org/10.1016/j.enpol.2013.12.050.

Opiyo, N. N. (2019). Impacts of neighbourhood influence on social acceptance of small solar home systems in rural western Kenya. *Energy Research & Social Science, 52*, 91–98. https://doi.org/10.1016/j.erss.2019.01.013.

O'Shaughnessy, E., Barbose, G., & Wiser, R. (2020). Patience is a virtue: A data-driven analysis of rooftop solar PV permitting timelines in the United States. *Energy Policy, 144*, 111615. https://doi.org/10.1016/j.enpol.2020.111615.

O'Shaughnessy, E., Grayson, A., & Barbose, G. (2023). The role of peer influence in rooftop solar adoption inequity in the United States. *Energy Economics, 127*, 107009. https://doi.org/10.1016/j.eneco.2023.107009.

Palm, A. (2017). Peer effects in residential solar photovoltaics adoption – A mixed methods study of Swedish users. *Energy Research & Social Science, 26*, 1–10. https://doi.org/10.1016/j.erss.2017.01.008.

Palm, A., & Lantz, B. (2020). Information dissemination and residential solar PV adoption rates: The effect of an information campaign in Sweden. *Energy Policy, 142*, 111540. https://doi.org/10.1016/j.enpol.2020.111540.

Papachristos, G. (2019). System dynamics modelling and simulation for sociotechnical transitions research. *Environmental Innovation and Societal Transitions, 31*, 248–261. https://doi.org/10.1016/j.eist.2018.10.001.

Parau, P., Lemnaru, C., Dinsoreanu, M., & Potolea, R. (2017). Opinion leader detection. In F. A. Pozzi, E. Fersini, E. Messina & B. Liu (Eds.) *Sentiment analysis in social networks* (pp. 157–170). Morgan Kaufmann. https://doi.org/10.1016/B978-0-12-804412-4.00010-3.

Parkins, J. R., Rollins, C., Anders, S., & Comeau, L. (2018). Predicting intention to adopt solar technology in Canada: The role of knowledge, public engagement, and visibility. *Energy Policy, 114*, 114–122. https://doi.org/10.1016/j.enpol.2017.11.050.

Pellicer-Sifres, V., Belda-Miquel, S., Cuesta-Fernández, I., & Boni, A. (2018). Learning, transformative action, and grassroots innovation: Insights from the Spanish energy cooperative Som Energia. *Energy Research & Social Science, 42*, 100–111. https://doi.org/10.1016/j.erss.2018.03.001.

Pisello, A. L., Castaldo, V. L., Piselli, C., Fabiani, C., & Cotana, F. (2016). How peers' personal attitudes affect indoor microclimate and energy need in an institutional building: Results from a continuous monitoring campaign in summer and winter conditions. *Energy and Buildings, 126*, 485–497. https://doi.org/10.1016/j.enbuild.2016.05.053.

Polski Alarm Smogowy (2020). *Razem o czyste powietrze. Raport z działalności w 2020 roku*. https://polskialarmsmogowy.pl/wp-content/uploads/2021/08/PAS_raport_2020.pdf.

Primc, K., & Slabe-Erker, R. (2020). Social policy or energy policy? Time to reconsider energy poverty policies. *Energy for Sustainable Development, 55*, 32–36. https://doi.org/10.1016/j.esd.2020.01.001.

Pudaruth, S., Juwaheer, T. D., & Koodruth, U. Y. (2017). Understanding the ecological adoption of solar water heaters among customers of island economies. *Studies in Business and Economics, 12*(1), 148–173. https://doi.org/10.1515/sbe-2017-0012.

Qing, C., He, J., Guo, S., Zhou, W., Deng, X., & Xu, D. (2022). Peer effects on the adoption of biogas in rural households of Sichuan Province, China. *Environmental Science and Pollution Research, 29*(40), 61488–61501. https://doi.org/10.1007/s11356-022-20232-y.

Rai, V., & Robinson, S. A. (2013). Effective information channels for reducing costs of environmentally-friendly technologies: evidence from residential PV markets. *Environmental Research Letters, 8*(1), 014044. https://doi.org/10.1088/1748-9326/8/1/014044.

Reed, M. S., Evely, A. C., Cundill, G., Fazey, I., Glass, J., Laing, A., Newig, J., Parrish, B., Prell, C., Raymond, C., & Stringer, L. C. (2010). What is social learning?. Ecology and society, 15(4).

Reuter, M., Narula, K., Patel, M. K., & Eichhammer, W. (2021). Linking energy efficiency indicators with policy evaluation – A combined top-down and bottom-up analysis of space heating consumption in residential buildings. *Energy and Buildings, 244*, 110987. https://doi.org/10.1016/j.enbuild.2021.110987.

Richter, L. L. (2013). Social effects in the diffusion of solar photovoltaic technology in the UK. *Cambridge Working Paper in Economics, 1357*. https://doi.org/10.17863/CAM.5680.

Rogers, E. M. (1962). *Bibliography on the Diffusion of Innovations*. Mimeo Bulletin AE 328.

Rogers, E. M. (1983). *Diffusion of Innovations. Third Edition.* A Division of Macmillan Publishing Co., Inc.

Scheller, F., Graupner, S., Edwards, J., Weinand, J., & Bruckner, T. (2022). Competent, trustworthy, and likeable? Exploring which peers influence photovoltaic adoption in Germany. *Energy Research & Social Science, 91*, 102755. https://doi.org/10.1016/j.erss.2022.102755.

Selvakkumaran, S., & Ahlgren, E. O. (2018). Model-Based Exploration of Co-Creation Efforts: The Case of Solar Photovoltaics (PV) in Skåne, Sweden. *Sustainability, 10*,(11), 3905. https://doi.org/10.3390/su10113905.

Shakeel, S.R., Rajala, A. (2020). Factors Influencing Households' Intention to Adopt Solar PV: A Systematic Review. In: Kantola, J., Nazir, S., Salminen, V. (eds) *Advances in Human Factors, Business Management and Leadership. AHFE 2020. Advances in Intelligent Systems and Computing*, vol 1209 (pp. 282–289). Springer, Cham. https://doi.org/10.1007/978-3-030-50791-6_36.

Shamon, H., Schumann, D., Fischer, W., Vögele, S., Heinrichs, H. U., & Kuckshinrichs, W. (2019). Changing attitudes and conflicting arguments: Reviewing stakeholder communication on electricity technologies in Germany. *Energy research & social science, 55*, 106–121. https://doi.org/10.1016/j.erss.2019.04.012.

Sinha, A., Sengupta, T., & Alvarado, R. (2020). Interplay between technological innovation and environmental quality: formulating the SDG policies for next 11 economies. *Journal of Cleaner Production, 242*, 118549. https://doi.org/10.1016/j.jclepro.2019.118549.

Spandagos, C., Baark, E., Ng, T. L., & Yarime, M. (2021). Social influence and economic intervention policies to save energy at home: Critical questions for the new decade and evidence from air-condition use. *Renewable and Sustainable Energy Reviews, 143*, 110915. https://doi.org/10.1016/j.rser.2021.110915.

Stefes, C. H. (2020). Opposing energy transitions: Modeling the contested nature of energy transitions in the electricity sector. *Review of Policy Research, 37*(3), 292–312. https://doi.org/10.1111/ropr.12381.

Sterman, J. D. (2000). *Business dynamics: Systems thinking and modeling for a complex world.* McGraw-Hill.

Sterman, J. D. (2001). System dynamics modeling: tools for learning in a complex world. *California management review, 43*(4), 8–25. https://doi.org/10.2307/41166098.

Tajfel, H., & Turner, J. C. (1979). An integrative theory of inter-group conflict. In W. G. Austin & S. Worchel (Eds.), *The social psychology of inter-group relations* (pp. 33–47). Monterey, CA: Brooks/Cole.

Tan, T. F., & Netessine, S. (2019). When you work with a superman, will you also fly? An empirical study of the impact of coworkers on performance. *Management Science, 65* (8), 3495–3517. https://doi.org/10.1287/mnsc.2018.3135.

Tippett, J., Searle, B., Pahl-Wostl, C., & Rees, Y. (2005). Social learning in public participation in river basin management – early findings from HarmoniCOP European case studies. *Environmental Science & Policy, 8*(3), 287–299. https://doi.org/10.1016/j.envsci.2005.03.003.

Trela, M., & Dubel, A. (2021). Net-metering vs. net-billing from the investors perspective – Impacts of changes in RES financing in Poland on the profitability of a joint photovoltaic panels and heat pump system. *Energies, 15*(1), 227. https://doi.org/10.3390/en15010227.

Valentine, V. (2010). *A narrative analysis of climate change coverage in "The New York Times", 1988–2008: Social responsibility and weight-of-evidence reporting.* Marquette University. http://epublications.marquette.edu/theses_open/33.

Van Est, R. (1999). *Winds of change: a comparative study of the politics of wind energy innovation in California and Denmark.* Utrecht: International books.

VanderWeele, T. J., & An, W. (2013). Social networks and causal inference. In S.L. Morgan (ed.). *Handbook of Causal Analysis for Social Research* (pp. 353–374). Springer.

Vibrans, L., Schulte, E., Morrissey, K., Bruckner, T., & Scheller, F. (2023). Same same, but different: explaining heterogeneity among potential photovoltaic adopters in Germany using milieu segmentation. *Energy Research & Social Science, 103*, 103212. https://doi.org/10.1016/j.erss.2023.103212.

Wang, Z. (2024). Strategic Insights into the Global Energy Landscape. *Highlights in Business, Economics and Management, 24*, 783–790. https://doi.org/10.54097/xvnc0625.

Wang, H., & Sun, B. (2022). New energy dominant technology diffusion mechanism based on cellular automation model – The case of the current market and policy environment. *International Journal of Energy Research, 46*(8), 10576–10589. https://doi.org/10.1002/er.7850.

Wenger, E. (2000). Communities of practice and social learning systems. *Organization, 7* (2), 225–246. https://doi.org/10.1177/135050840072002.

Wu, J., Guo, Y., Wu, M. Y., Morrison, A. M., & Ye, S. (2023). Green or red faces? Tourist strategies when encountering irresponsible environmental behavior. *Journal of Tourism and Cultural Change, 21*(4), 406–432. https://doi.org/10.1080/14766825.2022.2106789.

Zapata Riveros, J., Kubli, M., & Ulli-Beer, S. (2019). Prosumer communities as strategic allies for electric utilities: Exploring future decentralization trends in Switzerland. *Energy Research & Social Science, 57*, 101219. https://doi.org/10.1016/j.erss.2019.101219.

Zhang, J., Ballas, D., & Liu, X. (2023). Neighbourhood-level spatial determinants of residential solar photovoltaic adoption in the Netherlands. *Renewable Energy, 206*, 1239–1248. https://doi.org/10.1016/j.renene.2023.02.118.